Praise for *The Long Emergency:*

"Kunstler is America's version of an Old Testament prophet, a stinging social critic who warns of dark days ahead if we do not change the way we live."
—*Pulse*

"Funny, irreverent, and blunt . . . To his eternal credit, Kunstler doesn't predict the end of the world; he just doesn't think that Wal-Mart, monster homes or suburban high schools have much of a future."
—Andrew Nikiforuk, *The Globe and Mail*

"Kunstler displays a kind of macabre wit about the unpleasantness and strife that await us all. . . . His assertions have a neat way of doubling back to anticipate your critiques. If you express doubt about his views, then you may well be among the deluded masses too addicted to your McSUV and McSuburb to accept the reality that lies ahead."
—Katharine Mieszkowski, Salon.com

"James Howard Kunstler's *The Long Emergency* may be destined to become the Dante's *Inferno* of the twenty-first century. It graphically depicts the horrific punishments that lie ahead for Americans for more than a century of sinful consumption and sprawling communities, fueled by the profligate use of cheap oil and gas. Its central message—that the country will pay dearly unless it urgently develops new, sustainable community-scale food systems, energy sources, and living patterns—should be read, digested, and acted upon by every conscientious U.S. politician and citizen."
—Michael Shuman, author of
Going Local: Creating Self-Reliant Communities in a Global Age

"James Howard Kunstler has given us, with his usual engaging wit and verve, a new kind of post-apocalypse scenario. Instead of the nuclear or ice-age wasteland of our earlier imaginings, he has depicted with detailed extrapolation the civilization of the United States after the oil runs out and a great economic collapse occurs. It is a strangely Arcadian vision, like the agrarian America that Jefferson, Calhoun, and the Southern Agrarians dreamed of. But Kunstler has fleshed it out with delightful quirky insights and provided our science fiction writers with a fresh mise-en-scene."
—Frederick Turner, author of
The New World and *The Culture of Hope*

"An especial strength of this book is its break with some of the more pernicious strands in the contemporary left, specifically the left's knee-jerk rejection of America acting militarily in its national interest. . . . There are hints of Malthus here, and of Oswald Spangler's *Decline of the West* as well. Mr. Kunstler's book is a jeremiad, driven by authorial presence. Pithy, entertaining descriptions of historical phenomena like the Soviet Union . . . enliven the text, allowing the veteran commentator to expound on themes that might read leaden by a less facile wordsmith. . . . The book succeeds as an accessible primer to a looming crisis that could end the American way of life."

—A. G. Gancarski, *The Washington Times*

"In the annals of doomsday literature . . . *The Long Emergency* is destined to become the new standard. . . . Demands frank consideration of what up to now has been unthinkable: that the ascendancy of the human race might have been a temporary phenomenon . . . This case has been made before, but here it is made powerfully and articulately, with no apology and no hint of reprieve. . . . *The Long Emergency* represents a 'wake-up call' in the same sense that a hand grenade tossed through your bedroom window might serve as an alarm clock. The book is stark and frightening. Read it soon."

—Jim Charlier, *Daily Camera*

The Long Emergency

Books by the Author

NONFICTION

The City in Mind
Home from Nowhere
The Geography of Nowhere

NOVELS

Maggie Darling
Thunder Island
The Halloween Ball
The Hunt
Blood Solstice
An Embarrassment of Riches
The Life of Byron Jaynes
A Clown in the Moonlight
The Wampanaki Tales
World Made by Hand

The Long Emergency

Surviving the End of Oil, Climate Change, and Other Converging Catastrophes of the Twenty-First Century

James Howard Kunstler

Grove Press
New York

Published simultaneously in Canada
Printed in the United States of America

Library of Congress Cataloging-in-Publication Data
Kunstler, James Howard.
The long emergency : surviving the end of oil, climate change, and other converging catastrophes of the twenty-first century / James Howard Kunstler.
p. cm.
Includes bibliographical references.
ISBN-13: 978-0-8021-4249-8
1. Petroleum as fuel—United States—Social aspects. 2. Fossil fuels—United States—Social aspects. 3. Petroleum industry and trade—United States—History—20th century. 4. Petroleum industry and trade—United States—History—21st century. 5. Petroleum industry and trade—United States—Economic conditions—20th century. 6. Petroleum industry and trade—United States—Economic conditions—21st century. 7. Renewable energy sources—United States—Social aspects. 8. Climate and civilization. 9. Environmentalism—United States—Social aspects. I. Title.
TP355.K86 2005
303.4973—dc22 2001033464

Grove Press
an imprint of Grove/Atlantic, Inc.
841 Broadway
New York, NY 10003

Distributed by Publishers Group West
www.groveatlantic.com

10 11 12 13 14 16 15 14 13 12 11 10

This book is for my friends' children

Benjamin Golden

Norabelle Greenberger

Oliver and Nicky Edsforth

Contents

THE FATAL METAPHOR OF PROGRESS, WHICH MEANS LEAVING THINGS BEHIND US, HAS UTTERLY OBSCURED THE REAL IDEA OF GROWTH, WHICH MEANS LEAVING THINGS INSIDE US.

—G. K. Chesterton

DO THE GODS EXIST? I DO NOT KNOW, BUT THEY SURE ACT AS IF THEY DID.

—RF

One

Sleepwalking into the Future

Carl Jung, one of the fathers of psychology, famously remarked that "people cannot stand too much reality." What you're about to read may challenge your assumptions about the kind of world we live in, and especially the kind of world into which time and events are propelling us. We are in for a rough ride through uncharted territory.

It has been very hard for Americans—lost in dark raptures of nonstop infotainment, recreational shopping, and compulsive motoring—to make sense of the gathering forces that will fundamentally alter the terms of everyday life in technological society. Even after the terrorist attacks of September 11, 2001, that collapsed the twin towers of the World Trade Center and sliced through the Pentagon, America is are still sleepwalking into the future. We have walked out of our burning house and we are now headed off the edge of a cliff. Beyond that cliff is an abyss of economic and political disorder on a scale that no one has ever seen before. I call this coming time the Long Emergency.

What follows is a harsh view of the decades ahead and what will happen, in the United States. Throughout this book I will concern myself with what I believe *is* happening, what *will* happen, or what *is likely* to happen, not what I hope or wish will happen. This is an important distinction. It is my view, for instance, that in the decades to come the national government will prove to be so impotent and ineffective in managing the enormous vicissitudes we face that the United States may not survive as a nation in any meaningful sense but rather will devolve into a set of autonomous regions. I do not welcome a crack-up of our nation but I think it is a plausible outcome that we ought to be prepared to face. I have published several books critical of the suburban living arrangement, which I regard as deeply pernicious to our society. While I believe we will be better off living differently, I don't welcome the tremendous personal hardship that will result as the infrastructure of that life loses its value and utility. I predict that we are entering an era of titanic international military strife over resources, but I certainly don't relish the prospect of war.

If I hope for anything in this book, it is that the American public will wake up from its sleepwalk and act to defend the project of civilization. Even in the face of epochal discontinuity, there is a lot we can do to assure the refashioning of daily life around authentic local communities based on balanced local economies, purposeful activity, and a culture of ideas consistent with reality. It is imperative for citizens to be able to imagine a hopeful future, especially in times of maximum stress and change. I will spell out these strategies later in this book.

Our war against militant Islamic fundamentalism is only one element among an array of events already under way that will alter our relations with the rest of the world, and compel us to live differently at home—sooner rather than later—whether we like it or not. What's more, these world-altering forces, events, and changes will interact synergistically, mutually amplifying each other to accelerate and exacerbate the emergence of meta-problems. Americans are woefully unprepared for the Long Emergency.

Your Reality Check Is in the Mail

Above all, and most immediately, we face the end of the cheap fossil fuel era. It is no exaggeration to state that reliable supplies of cheap oil and natural gas underlie everything we identify as a benefit of modern life. All the necessities, comforts, luxuries, and miracles of our time—central heating, air conditioning, cars, airplanes, electric lighting, cheap clothing, recorded music, movies, supermarkets, power tools, hip replacement surgery, the national defense, you name it—owe their origins or continued existence in one way or another to cheap fossil fuel. Even our nuclear power plants ultimately depend on cheap oil and gas for all the procedures of construction, maintenance, and extracting and processing nuclear fuels. The blandishments of cheap oil and gas were so seductive, and induced such transports of mesmerizing contentment, that we ceased paying attention to the essential nature of these miraculous gifts from the earth: that they exist in finite, nonrenewable supplies, unevenly distributed around the world. To aggravate matters, the wonders of steady technological progress under the reign of oil have tricked us into a kind of "Jiminy Cricket syndrome," leading many Americans to believe that anything we wish for hard enough can come true. These days, even people in our culture who ought to know better are wishing ardently that a smooth, seamless transition from fossil fuels to their putative replacements—hydrogen, solar power, whatever—lies just a few years ahead. I will try to demonstrate that this is a dangerous fantasy. The true best-case scenario may be that some of these technologies will take decades to develop—meaning that we can expect an extremely turbulent interval between the end of cheap oil and whatever comes next. A more likely scenario is that new fuels and technologies may *never* replace fossil fuels at the scale, rate, and manner at which the world currently consumes them.

What is generally not comprehended about this predicament is that the developed world will begin to suffer long before the oil and gas actually run out. The American way of life—which is now virtually synonymous with suburbia—can run only on reliable supplies of dependably cheap oil and gas. Even mild to moderate deviations in either price or supply will crush our economy and make the logistics of daily life impossible. Fossil fuel reserves are not scattered equitably around the world. They tend to be concentrated in places where the native peoples don't like the West in general or America in particular, places physically very remote, places where we realistically can exercise little control (even if we wish to). For reasons I will spell out, we can be certain that the price and supplies of fossil fuels will suffer oscillations and disruptions in the period ahead that I am calling the Long Emergency.

The decline of fossil fuels is certain to ignite chronic strife between nations contesting the remaining supplies. These resource wars have already begun. There will be more of them. They are very likely to grind on and on for decades. They will only aggravate a situation that, in and of itself, could bring down civilizations. The extent of suffering in our country will certainly depend on how tenaciously we attempt to cling to obsolete habits, customs, and assumptions—for instance, how fiercely Americans decide to fight to maintain suburban lifestyles that simply cannot be rationalized any longer.

The public discussion of this issue has been amazingly lame in the face of America's post-9/11 exposure to the new global realities. As of this writing, no one in the upper echelon of the federal government has even ventured to state that we face fossil fuel depletion by mid-century and severe market disruptions long before that. The subject is too fraught with scary implications for our collective national behavior, most particularly the not-incidental fact that our economy these days is hopelessly tied to the creation and servicing of suburban sprawl.

Within the context of this feeble public discussion over our energy future, some wildly differing positions stand out. One faction of so-called "cornucopians" asserts that humankind's demonstrated technical ingenuity will overcome the facts of geology. (This would seem to be the default point of view of the majority of Americans, when they reflect on these issues at all.) Some cornucopians believe that oil is not fossilized, liquefied organic matter but rather a naturally occurring mineral substance that exists in endless abundance at the earth's deep interior like the creamy nougat center of a bonbon. Most of the public simply can't entertain the possibility that industrial civilization will not be rescued by technological innovation. The human saga has indeed been amazing. We have overcome tremendous obstacles. Our late-twentieth-century experience has been especially rich in technologic achievement (though the insidious diminishing returns are far less apparent). How could a nation that put men on the moon feel anything but a nearly godlike confidence in its ability to overcome difficulties?

The computer at which I am sitting would surely have been regarded as an astounding magical wonder by someone from an earlier period of American history, say Benjamin Franklin, who helped advance the early understanding of electricity. The sequence of discoveries and developments since 1780 that made computers possible is incredibly long and complex and includes concepts that we may take for granted, starting with 110-volt alternating house current that is always available. But what would Ben Franklin have made of video? Or software? Or broadband? Or plastic? By extension, one would have to admit the possibility that scientific marvels await in the future that would be difficult for people of our time to imagine. Humankind may indeed come up with some fantastic method for running civilization on seawater, or molecular organic nanomachines, or harnessing the *dark matter* of the universe. But I'd argue that such miracles may lie on the far shore of the Long Emergency, or may never happen at all. It is possible that the fossil fuel efflorescence was a one-shot deal for the human race.

A coherent, if extremely severe, view along these lines, and in opposition to the cornucopians, is embodied by the "die-off" crowd.[1] They believe that the carrying capacity of the planet has already exceeded "overshoot" and that we have entered an apocalyptic age presaging the imminent extinction of the human race. They lend zero credence to the cornucopian belief in humankind's godlike ingenuity at overcoming problems. They espouse an economics of net entropy. They view the end of oil as the end of everything. Their worldview is terminal and tragic.

The view I offer places me somewhere between these two camps, but probably a few degrees off center and closer to the die-off crowd. I believe that we face a dire and unprecedented period of difficulty in the twenty-first century, but that humankind will survive and continue further into the future—though not without taking some severe losses in the meantime, in population, in life expectancies, in standards of living, in the retention of knowledge and technology, and in decent behavior. I believe we will see a dramatic die-back, but not a die-off. It seems to me that the pattern of human existence involves long cycles of expansion and contraction, success and failure, light and darkness, brilliance and stupidity, and that it is grandiose to assert that our time is so special as to be the end of all cycles (though it would also be consistent with the narcissism of baby-boomer intellectuals to imagine ourselves to be so special). So I have to leave room for the possibility that we humans will manage to carry on, even if we must go through this dark passage to do it. We've been there before.

The Groaning Multitudes

It has been estimated that the world human population stood at about one billion around the early 1800s, which was roughly about when the

1. *www.dieoff.com,* an Internet site started by Jay Hanson, popularizing the ideas of many who believe that the Industrial Age is a terminal condition of humankind.

industrial adventure began to gain traction.[2] It has been inferred from this that a billion people is about the limit that the planet Earth can support when it is run on a nonindustrial basis. World population is now past six and a half billion, having more than doubled since my childhood in the 1950s. The mid-twentieth century was a time of rising anxiety over the "population explosion." The marvelous technological victory over food shortages, including the "green revolution" in crop yields, accelerated that already robust leap in world population that had begun with modernity. Dramatic improvements in sanitation and medicine extended lives. Industry sopped up expanding populations and reassigned them from rural lands to work in the burgeoning cities. The perceived ability of the world to accommodate these newcomers and latecomers in a wholly new disposition of social and economic arrangements seemed be the final nail in the coffin of Thomas Robert Malthus, the much-abused author of the 1798 "An Essay on the Principle of Population as It Affects the Future Improvement of Society."

Malthus (1766–1834), an English country clergyman educated at Cambridge, has been the whipping boy of idealists and techno-optimists for two hundred years. His famous essay proposed that human population, if unconstrained, would grow exponentially while food supplies grew only arithmetically, and that therefore population growth faced strict and inevitable natural limits. Most commentators, however, took the math at face value and overlooked the part about constraints. These "checks" on population come in the form of famine, pestilence, war, and "moral restraint," i.e., the will to postpone marriage or forgo parenthood (from a perhaps antiquated notion that the ability to support a family might enter into anyone's plans for forming one, or even that society could influence such choices). Malthus's essay has been mostly misconstrued to mean that the human race was doomed at a certain arbitrary set point, and the pejorative "Malthusian" is attached to any idea that suggests that human ingenuity cannot make accommodation for more human beings to join the party on Spaceship Earth.

2. Historian Paul Johnson's notion of "the Modern" commencing around the end of the Napoleonic Wars is a good benchmark. See Johnson, *The Birth of the Modern*, New York: Harper, 1991.

Interestingly, Malthus's essay was aimed at the reigning Enlightenment idealists of his own youth, the period of the American and French Revolutions, in particular the seminal figures of William Godwin and the Marquis de Condorcet. Both held that mankind was infinitely improvable and that a golden age of social justice, political harmony, equality, abundance, brotherhood, happiness, and altruism loomed imminently. Although sympathetic to social improvement, Malthus deemed these claims untenable and thought it necessary to debunk them.

In recent times, population pessimists such as Paul Ehrlich, author of *The Population Bomb* (1968), Lester Brown of the Worldwatch Institute, and other commentators who predicted dire consequences of overpopulation by 1980, were supposedly shown up by the failure of dire events to occur; this led a new generation of idealists (including cornucopians such as economist Julian Simon) to proclaim that hypergrowth was a positive benefit to society because the enlarged pool of social capital and intellect would inevitably lead to fantastic new technological discoveries that would in turn permit the earth to support a greater number of humans—including social or medical innovations that would aid eventually in establishing a permanently stabilized optimum human population.

I would offer a different view. Malthus was certainly correct, but cheap oil has skewed the equation over the past hundred years while the human race has enjoyed an unprecedented orgy of nonrenewable condensed solar energy accumulated over eons of prehistory. The "green revolution" in boosting crop yields was minimally about scientific innovation in crop genetics and mostly about dumping massive amounts of fertilizers and pesticides made out of fossil fuels onto crops, as well as employing irrigation at a fantastic scale made possible by abundant oil and gas. The cheap oil age created an artificial bubble of plenitude for a period not much longer than a human lifetime, a hundred years. Within that comfortable bubble the idea took hold that only grouches, spoilsports, and godless maniacs considered population hypergrowth a problem, and that to even raise the issue was indecent. So, I hazard to assert that as oil ceases to be cheap and the world reserves arc toward depletion, we will indeed suddenly be left with an enormous surplus population—with apologies to both Charles Dickens and Jonathan Swift—that the ecology of the earth will not support. No political program of birth control will avail. The people are already here. The journey back to non-oil population homeostasis will not be pretty. We will discover the hard way that population hypergrowth was simply a side effect of the oil age. It was a condition, not a problem with a solution. That is what happened and we are stuck with it.

Trashed Planet

We are already experiencing huge cost externalities from population hypergrowth and profligate fossil fuel use in the form of environmental devastation. Of the earth's estimated 10 million species, 300,000 have vanished in the past fifty years. Each year, 3,000 to 30,000 species become extinct, an all-time high for the last 65 million years. Within one hundred years, between one-third and two-thirds of all birds, animals, plants, and other species will be lost. Nearly 25 percent of the 4,630 known mammal species are now threatened with extinction, along with 34 percent of fish, 25 percent of amphibians, 20 percent of reptiles, and 11 percent of birds. Even more species are having population declines.[3] Environmental scientists speak of an "omega point" at which the vast interconnected networks of Earth's ecologies are so weakened that human existence is no longer possible. This is a variant of the die-off theme that I consider unlikely, but it does raise grave questions about the ongoing project of civilization. How long might the Long Emergency last? A generation? Ten generations? A millennium? Ten millennia? Take your choice. Of course, after a while, an emergency becomes the norm and is no longer an emergency.

Global warming is no longer a theory being disputed by political interests, but an established scientific consensus.[4] The possible effects

3. *World Watch*, Jan./Feb. 1997, p. 7.

4. Authorities who agree that global warming and climate change are real and serious problems include the National Academy of Sciences, the UN-sponsored World Meteorological Association's Intergovernmental Panel on Climate Change (IPCC), the National Oceanic and Atmospheric Administration, the U.S. Environmental Protection Agency, the U.S. Department of Energy, NASA's Goddard Institute, the Union of Concerned Scientists, the World Resources Institute, and many others.

range from events as drastic as a hydrothermal shutdown of the Gulf Stream—meaning a much colder Europe with much reduced agriculture—to desertification of major world crop-growing areas, to the invasion of temperate regions by diseases formerly limited to the tropics, to the loss of harbor cities all over the world. Whether the cause of global warming is human activity and "greenhouse emissions," a result of naturally occurring cycles, or a combination of the two, this does not alter the fact that it is having swift and tremendous impacts on civilization and that its effects will contribute greatly to the Long Emergency.

Global warming projections by the Intergovernmental Panel on Climate Change (IPCC) show a widespread increase in the risk of flooding for tens of millions of people due to increased storms and sea-level rise. Climate change is projected to aggravate water scarcity in many regions where it is already a problem. It will increase the number of people exposed to vector-borne disease (e.g., malaria and dengue fever) and waterborne disease (e.g., cholera). It will obviate the triumphs of the green revolution and bring on famines. It will prompt movements of populations fleeing devastated and depleted lands and provoke armed conflicts over places that are better endowed.

Global warming will add a layer of further desperation to the political turmoil ensuing from contests over dwindling oil supplies. It will aggravate the environmental destruction in China, where massive desertification and freshwater depletion are already at crisis levels, in a nation grossly overpopulated and attempting to industrialize just as the means for industrializing worldwide are diminishing. Global warming will contribute to conditions that will shut down the global economy.

Revenge of the Rain Forest and Other Tiny Destroyers

The high tide of the cheap oil age also happened to be a moment in history when human ingenuity gained an upper hand against the age-old scourges of disease. We have enjoyed the great benefits of antibiotic medicine for roughly a half-century. Penicillin, sulfa drugs, and their descendants briefly gave mankind the notion that diseases caused by microorganisms could, and indeed would, be systematically vanquished. Or, at least, this was the popular view. Doctors and scientists knew better. The discoverer of penicillin, Alexander Fleming, himself warned that antibiotic misuse could result in resistant strains of bacteria.

The recognition is now growing that the victory over microbes was short-lived. They are back in force, including familiar old enemies such as tuberculosis and staphylococcus in new drug-resistant strains. Other old diseases are on the march into new territories, as a response to climate change brought on by global warming. In response to unprecedented habitat destruction by humans, and the invasion of wilderness, the earth itself seems to be sending forth new and much more lethal diseases, as though it had a kind of protective immune system with antibodylike agents aimed with remarkable precision at the source of the problem: *Homo sapiens*. Human immunodeficiency virus (HIV), the precursor of AIDS, may be the revenge of the rain forest. In the twentieth century, a critical mass of humans encountered organisms long hidden in tropical backwaters, presenting ripe targets for opportunistic mutant strains of immunodeficiency virus jumping species. Once infected, these humans are able to travel out of the rain forest, courtesy of motor vehicles, and reenter the social mainstream with a newly acquired ability to infect others. One theory holds that HIV first developed in the 1940s from the simian immunodeficiency virus (SIV), which has long infected green monkeys, mangabeys, and baboons in Africa. The human immunodeficiency viruses HIV-2 and HIV-O both bear similarities to SIV. The virus may have jumped to humans through the consumption of monkeys as so-called "bush meat," or through monkey bites. The virus may have infected human hosts, where it then mutated into its current lethal form. HIV probably first infected rural areas of Africa, slowly moving into the cities and around the world until it hit homosexual communities, where conditions were sufficient for rapid transmission of the disease via blood and, incidentally, other body fluids. AIDS also enjoyed the advantage of having a long incubation period so that in the initial stage of the epidemic, few if any carriers had any idea that they harbored a vicious disease, allowing them to unwittingly spread the disease further.

In any case, AIDS is now a growing menace—despite the illusion in wealthy nations such as the United States that it is a manageable chronic illness—with its cases doubling every two years around the world. Having exploded across sub-Saharan Africa, it is now marching with increasing lethality into the most heavily populated parts of the world: India and China. The virus mutates continually and there may be variants too numerous to count. It has been transmitted through homosexual and heterosexual acts, by needle sharing among intravenous drug addicts, and lately in China among commercial blood harvesters reusing needles. The virus could hardly have exhausted its ability to mutate into new modes of transmission, and while that ought to be a big worry for all human societies, there is probably little that can be done about it. A deadly emergent system has been set in motion and it has not finished emerging. All other human problems may pale in comparison to the AIDS epidemic in another ten years if infection rates continue along their current arc.

At the same time, the world is overdue for an extreme influenza epidemic. The last major outbreak was the 1918 Spanish influenza, which killed 50 million people worldwide and changed the course of history. That flu, which seems to have originated on a Kansas pig farm, affected the outcome of World War I, toppled three dynasties (the Hohenzollerns in Germany, the Hapsburgs of Austria, and the Romanovs of Russia), and set the course of the world toward fascism, communism, and the Second World War.

Disease will certainly play a larger role in the Long Emergency than many can now imagine. An epidemic could paralyze social and economic systems, interrupt global trade, and bring down governments. Regimes overwhelmed with population pressures—at a time of crashing worldwide oil supplies and a melting global economic system—might be tempted to deploy "designer" viruses against their own masses, inoculating beforehand an elite of select survivors. Disease would provide a convenient moral cover for an act of political desperation. The medical technology is certainly available. If this sounds too fantastic, imagine how outlandish the liquidation of European Jewry might have seemed to civilized Berliners in 1933. Yet it happened. The machinery of the Holocaust employed all the latest state-of-the-art industrial technology, and it was carried out by the statistically best-educated nation in Europe.

At the very least, the Long Emergency will be a time of diminished life spans for many of us, as well as reduced standards of living—at least as understood within the current social context. Fossil fuels had the effect of temporarily raising the carrying capacity of the earth. Our ability to resist the environmental corrective of disease will probably prove to have been another temporary boon of the cheap-oil age, like air conditioning and lobsters flown daily from Maine to the buffets of Las Vegas. So much of what we construe to be among our entitlements to perpetual progress may prove to have been a strange, marvelous, and anomalous moment in the planet's history.

Adios Globalism

The so-called global economy was not a permanent institution, as some seem to believe it was, but a set of transient circumstances peculiar to a certain time: the Indian summer of the fossil fuel era. The primary enabling mechanism was a world-scaled oil market allocation system able to operate in an extraordinary sustained period of relative world peace. Cheap oil, available everywhere, along with ubiquitous machines for making other machines, neutralized many former comparative advantages, especially of geography, while radically creating new ones—hypercheap labor, for instance. It no longer mattered if a nation was halfway around the globe, or had no prior experience with manufacturing. Cheap oil brought electricity to distant parts of the world where ancient traditional societies had previously depended on renewables such as wood and dung, mainly for cooking, as many of these places were tropical and heating was not an issue. Factories could be started up in Sri Lanka and Malaysia, where swollen populations furnished trainable workers willing to labor for much less than those back in the United States or Europe. Products then moved around the globe in a highly rationalized system, not unlike the oil allocation system, using immense vessels, automated port facilities, and truck-scaled shipping containers at a minuscule cost-per-unit of whatever was made and transported. Shirts or coffeemakers manufactured 12,000 miles away could be shipped to Wal-Marts all over America and sold cheaply.

The ability to globalize industrial manufacturing this way stimulated a worldwide movement to relax trade barriers that had existed previously to fortify earlier comparative advantages, which were now deemed obsolete. The idea was that a rising tide of increased world trade would lift all boats. The period (roughly 1980–2001) during which these international treaties relaxing trade barriers were made—the General Agreements on Tariffs and Trade (GATT)—coincided with a steep and persistent drop in world oil and gas prices that occurred precisely because the oil crises of the 1970s had stimulated so much frantic drilling and extraction that a twenty-year oil glut ensued. The glut, in turn, allowed world leaders to forget that the globalism they were engineering depended wholly on nonrenewable fossil fuels and the fragile political arrangements that allowed their distribution. The silly idea took hold among the free, civilized people of the West, and their leaders, that the 1970s oil crises had been fake emergencies, and that oil was now actually superabundant. This was a misunderstanding of the simple fact that the North Sea and Alaskan North Slope oil fields had temporarily saved the industrial West when they came online in the early 1980s, and postponed the fossil fuel depletion reckoning toward which the world has been inexorably moving.

Meanwhile, among economists and government figures, globalism developed the sexy glow of an intellectual fad. Globalism allowed them to believe that burgeoning wealth in the developed countries, and the spread of industrial activity to formerly primitive regions, was based on the potency of their own ideas and policies rather than on cheap oil. Margaret Thatcher's apparent success in turning around England's sclerotic economy was an advertisement for these policies, which included a heavy dose of privatization and deregulation. Overlooked is that Thatcher's success in reviving England coincided with a fantastic new revenue stream from North Sea oil, as quaint old Britannia became energy self-sufficient and a net energy-exporting nation for the first time since the heyday of coal. Globalism then infected America when Ronald Reagan came on the scene in 1981. Reagan's "supply-side" economic advisors retailed a set of fiscal ideas that neatly accessorized the new notions about free trade and deregulation, chiefly that massively reducing taxes would actually result in greater revenues as the greater aggregate of business activity generated a greater aggregate of taxes even at lower rates. (What it actually generated was huge government deficits.)

By the mid-1980s deregulated markets and unbridled business were regarded as magic bullets to cure the ills of senile smokestack industrialism. Greed was good. Young college graduates marched into MBA programs in hordes, hoping to emerge as corporate ninja warriors. It was precisely the entrepreneurial zest of brilliant young corporate innovators that produced the wizardry of the computer industry. The rise of computers, in turn, promoted the fantasy that commerce in sheer information would be the long-sought replacement for all the played-out activities of the smokestack economy. A country like America, it was now thought, no longer needed steelmaking or tire factories or other harsh, dirty, troublesome enterprises. Let the poor masses of Asia and South America have them and lift themselves up from agricultural peonage. America would outsource all this old economy stuff and use computers to orchestrate the movement of parts and the assembly of products from distant quarters of the world, and then sell the stuff in our own Kmarts and Wal-Marts, which would become global juggernauts of retailing. Computers, it was believed, would stupendously increase productivity all the way down the line. The jettisoned occupational niches in industry would be replaced by roles in the service economy that went hand in hand with the information economy. We would become a nation of hair stylists, masseurs, croupiers, restaurant owners, and show business agents, catering to one another's needs. Who wanted to work in a rolling mill?

Finally, the disgrace of Soviet communism in the early 1990s resolved any lingering philosophical complaints among the educated classes about the morality of business per se and of the institutions needed to run it. The Soviet fiasco had proven that a state without property laws or banking was just another kind of scaled-up social Ponzi scheme running on cheap oil and slave labor.[5]

5. Charles K. Ponzi (1888–1949), an Italian-born swindler who emigrated first to Canada, in 1903, where he served a prison term for forgery, came to the United States in 1920 and devised an investment fraud based on the same "pyramid" principles as a chain letter. Early investors in the postal coupon scam, which Ponzi ran out of Boston, were paid with money from later investors. Within six months, Ponzi was in the hands of federal prosecutors. He was sentenced to five years on a single plea-bargained count of mail fraud. He later engaged in Florida land swindles. He spent the 1930s in and out of state and federal penitentiaries. He was eventually deported back to Italy, moved to Brazil before World War II, and died in the charity ward of a Rio de Janeiro hospital in 1949, leaving an estate of $75 to cover funeral expenses.

In the short term, finance also benefited hugely from the removal of legal barriers to trade in currencies and investment instruments between nations. Computers enabled money to move around the planet at the speed of light. Investors in Luxembourg could just as easily invest in American securities, or in China's, as in their own. Other players benefited from trading in world currencies, securities, commodities, and interest rates at minute differentials that existed because, since the 1970s, all monies and fungible financial instruments pegged to money floated on a collective hallucination of relative value, rather than being pegged to a fixed medium of value, such as gold. This aggravated the tendency, in a financial climate of extreme relativism, to create increasingly abstract vehicles of investment that were pegged to little more than wishes. These so-called derivatives ended up far removed from the actual purpose of investment, which is to pay for new or expanded enterprise in return for earnings and dividends, and instead simply became an end in themselves: bets within global finance casinos. Eventually, this speculative trade was carried on by firms and individuals at such huge increments that whole national currencies and economies could be undermined, as when financier George Soros devalued the British pound in a single currency trade gamble, or when the Long-Term Capital Management firm, operating out of a luxury boiler room in suburban Connecticut, nearly destabilized the entire world finance system in a skein of fantastically huge and complex hedged derivatives trades—i.e., wild bets.

Finance under globalism, or turbo-capitalism (Edward Luttwak's term), or neoliberal economics (John Gray's term) took on the characteristics of a worldwide pyramid racket, played against the background of a geopolitical game of musical chairs.[6] In this case, the profits of a generation of speculators would be converted into costs passed along to future generations in the form of lost jobs, squandered equity, and reduced living standards. It was also like a convoluted liquidation sale of the accrued wealth of two hundred years of industrial society for the benefit of a handful of financial buccaneers, with the great masses relegated to a race to

6. Edward Luttwak, *Turbo Capitalism: Winners and Losers in the Global Economy*, New York: HarperCollins, 1999; John Gray, *False Dawn: The Delusions of Global Capitalism*, New York: New Press, 1998.

the bottom as the economic assets are dismantled and sold off, and their livelihoods are closed down. Both Luttwak and Gray make the case that millennial economics produced ever-greater disparities between winners and losers, between the wealthy and the poor, and that these deformities of economic behavior have the power to wreck societies.

I have argued in previous books that capitalism is not strictly speaking an "ism," in the sense that it is not so much a set of beliefs as a set of laws describing the behavior of money as it relates to accumulated real wealth or resources. This wealth can be directed toward the project of creating more wealth, which we call investment, and the process can be rationally organized within a body of contract and property law. Within that system are many subsets of rules and laws that describe the way money in motion operates, much as the laws of physics describe the behavior of objects in motion. Concepts such as interest, credit, revenue, profit, and default don't require a belief in capitalism in order to operate. Compound interest has worked equally well for communists and Wall Street financiers, whatever they personally thought about the social effects of wealth and poverty. People of widely differing beliefs are also equally subject to the law of gravity.

It is therefore not a matter of whether people believe in capitalism (hyper, turbo, neoliberal, or anything else you might call it), but of the choices they make as individuals, and in the aggregate as communities and nations, that determine their destiny. I am going to argue in later chapters that Americans in particular among the so-called "advanced" nations made some especially bad choices as to how they would behave in the twilight of the fossil fuel age. For instance, conditions over the past two decades made possible the consolidation of retail trade by a handful of predatory, opportunistic corporations, of which Wal-Mart is arguably the epitome. That this development was uniformly greeted as a public good by the vast majority of Americans, at the same time that their local economies were being destroyed—and with them, myriad social and civic benefits—is one of the greater enigmas of recent social history. In effect, Americans threw away their communities in order to save a few dollars on hair dryers and plastic food storage tubs, never stopping to reflect on what they were destroying. The necessary restoration of local networks of economic interdependence, and the communities that rely on them, will be a major theme later in this book.

I will also propose that globalism as we have known it is in the process of ending. Its demise will coincide with the end of the cheap-oil age. For better or worse, many of the circumstances we associate with globalism will be reversed. Markets will close as political turbulence and military mischief interrupt trade relations. As markets close, societies will turn increasingly to import replacement for sheer economic survival. The cost of transport will no longer be negligible in a post-cheap-oil age. Many of our agricultural products will have to be produced closer to home, and probably by more intensive hand labor as oil and natural gas supplies become increasingly unstable. The world will stop shrinking and become larger again. Virtually all of the economic relationships among persons, nations, institutions, and things that we have taken for granted as permanent will be radically changed during the Long Emergency. Life will become intensely and increasingly local.

The End of the Drive-In Utopia

America finds itself nearing the end of the cheap-oil age having invested its national wealth in a living arrangement—suburban sprawl—that has no future. When media commentators cast about struggling to explain what has happened in our country economically, they uniformly overlook the colossal misinvestment that suburbia represents—a prodigious, unparalleled misallocation of resources. This is quite apart from its social, spiritual, and ecological deficiencies as an everyday environment. We constructed an armature for daily living that simply won't work without liberal supplies of cheap oil, and very soon we will be without both the oil needed to run it and the wealth needed to replace it. Nor are we likely to come up with a miraculous energy replacement for oil that will allow us to run all this everyday infrastructure even remotely the same way. I will go into detail about the mirage of alternative fuels later, in Chapter Four.

In any case, the tragic truth is that much of suburbia is unreformable. It does not lend itself to being retrofitted into the kind of mixed-use, smaller-scaled, more fine-grained walkable environments we will need to carry on daily life in the coming age of greatly reduced motoring. Nor is a Jolly Green Giant going to come and pick up the millions of suburban houses on their half-acre lots on cul-de-sac streets in the far-flung subdivisions and set them back down closer together to make more civic environments. Instead, this suburban real estate, including the chipboard and vinyl McHouses, the strip malls, the office parks, and all the other components, will enter a phase of rapid and cruel devaluation. Many of the suburban subdivisions will become the slums of the future.

Overall, I view the period ahead as one of generalized and chronic contraction. In the final chapter I will discuss comprehensively what this means in terms of how we may have to live. I refer to this process as the downscaling of America—rescaling or rightsizing might be other ways to say it. All of our accustomed modes of activity are going to have to change in the direction of smaller, fewer, and better. The crisis in agriculture will be one of the defining conditions of the Long Emergency. We will simply have to grow more of our food locally. The crisis will present itself when industrial farming, dependent on massive oil and gas "inputs" at gigantic scales of operation, can no longer be carried on economically. The implications for how we use our land are tremendous, and the unavoidable change is likely to be accompanied by severe social turbulence, not to mention hunger and hardship. Well into the Long Emergency, food production at the local level may become the focus of the American economy. The fact that it will almost certainly require a lot of human labor has further implications of its own.

We'll have to live in geographically more circumscribed surroundings. As the suburbs disintegrate, we will be lucky if we can reconstitute our existing traditional towns and cities brick by brick and street by street, painfully by hand. Our bigger cities will be in trouble, and some of them may not remain habitable, especially if the natural gas supply problem proves to be as dire as it now appears and electric power generation that depends on it becomes erratic. Skyscrapers will prove to be more experimental than we had come to think. In general, we will probably have to return to a settlement pattern of towns and small cities surrounded by intensively cultivated agricultural hinterlands. When that happens, we will be a far less affluent society and the amount, scale, and increment of new building will seem very modest in years ahead by current standards. We will have access to far fewer, if any, modular building systems. Construction will be much more dependent on traditional masonry, carpentry, and other journeyman skills using simple, easily obtainable, regionally determined materials. Our building and zoning codes will be increasingly ignored. If we return to a human scale of building, there's a good chance that our new urban quarters will be more humane, which is to say beautiful. The automobile era proved that people easily tolerated ugly, utilitarian buildings and horrible streetscapes as long as they were compensated by being able to quickly escape the vicinity in cars luxuriously appointed with the finest digital stereo sound, air conditioning, and cup holders for iced beverages. This will change radically. There will be far less motoring. The future will be much more about staying where you are than traveling incessantly from place to place, as we do now.

The state-of-the-art mega-suburbs of recent decades have produced horrendous levels of alienation, loneliness, anomie, anxiety, and depression, and we may well be better off without them. Note, by the way, that we have been the only nation among the so-called advanced ones to sacrifice our traditional cities and towns so remorselessly to suburbia. Elsewhere, in Europe, Asia, and South America, whatever else their problems may be, cities and towns still exist intact in a more distinct relationship with nearby rural lands. The restoration job in America will be more difficult.

But since I believe that the human race will carry on for many generations after the end of the cheap-oil era, and that civilization of some form can continue with it, then I would have to suppose that the seasons of civilization will continue with the great cycles of contraction and expansion, and at some point in the future, who knows how many years distant, some of these cities in a land once called America may be robust and cosmopolitan in ways that we can't imagine now, anymore than a Roman of A.D. 38 might have been able to imagine the future London of the Beatles.

In the Long Emergency, some regions of the United States will do better than others and some will suffer deeply. Places that benefited disproportionately during the cheap-oil blowout will find themselves steeply challenged when those benefits, and the entitlement psychology that grew out of them, are withdrawn in the face of new, austere circumstances. The so-called Sunbelt presents extraordinary problems. This is not a good time to begin thinking about moving to Phoenix or Las Vegas. Parts of the Southwest may be significantly depopulated, starved for energy and thirsting for water that depended on cheap energy. Other parts may become contested territory with Mexico. The prospects for disorder in the southeastern states is especially high, given the extremes of religiosity, hyperindividualism, and a cultural disinhibition regarding violence. The social glue holding communities and regions together will be severely strained by the loss of amenities presumed to be normal.

I view the period of history we have lived through as a narrative episode in a greater saga of human history. The industrial story has a beginning, a middle, and an end. It begins in the mid-eighteenth century with coal and the first steam engines, proceeds to a robust second act climaxing in the years before World War I, and moves toward a third act resolution now that we can anticipate with some precision the depletion of the resources that made the industrial episode possible. As the industrial story ends, the greater saga of mankind will move on into a new episode, the Long Emergency. This is perhaps a self-evident point, but throughout history, even the most important and self-evident trends are often completely ignored because the changes they foreshadow are simply unthinkable. That process is sometimes referred to as an "outside context problem," something so far beyond the ordinary experience of those dwelling in a certain time and place that they cannot make sense of available information. The collective mental static preventing comprehension is also sometimes referred to as "cognitive dissonance," a term borrowed from developmental psychology. It helps explain why the American public has been sleepwalking into the future.

The Long Emergency is going to be a tremendous trauma for the human race. It is likely to entail political turbulence every bit as extreme as the economic conditions that prompt it. We will not believe that this is happening to us, that two hundred years of modernity can be brought to its knees by a worldwide power shortage. The prospect will be so grim that some individuals and perhaps even groups (as in nations) may develop all the symptoms of suicidal depression. Self-genocide has certainly been within the means of mankind since the 1950s.

The survivors will have to cultivate a religion of hope, that is, a deep and comprehensive belief that humanity is worth carrying on. I say this as someone who has not followed any kind of lifelong organized religion. But I don't doubt that the hardships of the future will draw even the most secular spirits into an emergent spiritual practice of some kind. There is an excellent chance that this will go way too far, as Christianity and other belief systems have done at various times, in various ways.

If it happens that the human race doesn't make it, then the fact that we were here once will not be altered, that once upon a time we peopled this astonishing blue planet, and wondered intelligently at everything about it and the other things who lived here with us on it, and that we celebrated the beauty of it in music and art, architecture, literature, and dance, and that there were times when we approached something godlike in our abilities and aspirations. We emerged out of depthless mystery, and back into mystery we returned, and in the end the mystery is all there is.

Two

Modernity and the Fossil Fuels Dilemma

The radio, and the telephone, and the movies that we know
May just be passing fancies, and in time may go.

—George and Ira Gershwin

In idle moments, I try to amuse myself by projecting my mind into other historical periods. Lately I am fascinated by what it must have been like to live in the early twentieth century when so many of the things we take for granted in our daily doings today had just come on the scene and established themselves as normal accessories to everyday life—the car, airplanes, household electricity, central heating, skyscrapers, radio, motion pictures, hot water on demand, X-rays. How *modern* it all must have seemed in 1924, when most adults could still remember a world of horse-drawn carriages, outhouses, kerosene lamps, and Saturday night baths! Whole ideologies had to be constructed to account for being modern and to explain it. It was celebrated in popular song.

Mostly everything that followed in applied technology was just a refinement to some degree of these original miracles, as TV was a refinement of radio, and magnetic resonance imaging was of the X-ray picture. How amazing it must have been to witness everyday life improving so dramatically, and how this procession of marvels must have induced people to think that the human race was moving toward exactly the sort of perfection that the Enlightenment philosophers had promised. The most astonishing thing, though, is how quickly we came to take these things for granted.

For thousands of years human beings dreamed rhapsodically of flying like birds. It was a persistent dream, both archetypal and unattainable.

The very wish to fly became synonymous with the defying of gods (and was suitably punished). The great minds of the ages, like Leonardo da Vinci, tried and failed to figure out some mechanical means to accomplish flight. Then, along came gasoline and the internal combustion engine and, voilà, within a few years a couple of fairly ordinary young bicycle mechanics from Ohio made it happen. (If the Wright brothers even had an affectionate name for their pioneering aircraft it is lost to history; such was the scope of their imaginations. The banality of American exceptionalism is sometimes astounding.) Now, exactly a hundred years after the first powered flight at Kitty Hawk, North Carolina, I can get on a jet airplane twice the size of a house several times a month and fly halfway across North America in the time it takes to finish a newspaper—and I end up feeling cranky and resentful about the service, to boot! They ran out of pretzels! The air conditioning was set too low!

A mere sixty-six years after the Wright brothers got airborne in their clunky, kitelike rig, the U.S. government flew men to the moon and back. (They played golf there.) Who even thinks about them anymore? (Lesson: Even "magic" has diminishing returns.)

Everything characteristic about the condition we call modern life has been a direct result of our access to abundant supplies of cheap fossil fuels. Fossil fuels have permitted us to fly, to go where we want to go rapidly, and move things easily from place to place. Fossil fuels rescued us from the despotic darkness of the night. They have made the pharaonic scale of building commonplace everywhere. They have allowed a fractionally tiny percentage of our swollen populations to produce massive amounts of food. They have allowed us to develop industries of surpassing ingenuity and to push the limits of what it even means to be human to the strange frontier where man imagines himself into a kind of machine immortality.

All of the marvels and miracles of the twentieth century were enabled by our access to abundant supplies of cheap fossil fuels. Even the applied technology of atomic fission, which came along at mid-century, would have been impossible without fossil fuels, and may be impossible to continue very long into the future without them.

The age of fossil fuels is about to end. There is no replacement for them at hand. These facts are poorly understood by the global population preoccupied with the thrum of daily life, but tragically, too, by the educated classes in the United States, who continue to be by far the greatest squanderers of fossil fuels. It is extremely important that we make an effort to understand what is about to happen to us because it will have earth-shaking repercussions for the way we live, the way the world is ordered, and on whether the very precious cargo of human culture can move safely forward into the future.

Global Peak

The key to understanding what is about to happen to us is contained in the concept of the *global oil production peak*. This is the point at which we have extracted half of all the oil that has ever existed in the world—the half that was easiest to get, the half that was most economically obtained, the half that was the highest quality and cheapest to refine. The remaining oil is the stuff that lies in forbidding places not easily accessed, such as the Arctic and deep under the ocean. Much of the remaining half is difficult to extract and may, in fact, take so much energy to extract that it is not worth getting—for instance, if it takes a barrel of oil to get a barrel of oil out of the ground, then you are engaged in an act of futility. If it takes two barrels of oil to get one barrel of oil, then you are engaged in an act of madness. Much of the remaining half comes in the form of high-sulfur crude, which is difficult to refine, or tar sands and oil shales, which are not liquids but solids that must be mined before they are liquefied for refining, adding two additional layers of expense on their recovery. Quite a bit of the remaining half of the world's original oil supply will never be recovered.

To move beyond the world oil production peak means that never again will all the nations of the earth combined extract as much oil from the ground as we did at peak, no matter what happens on the demand side. This has extraordinary implications for oil-based industrial civilization, which is predicated on constant and regular expansion of everything—population, gross domestic product, sales, revenue, housing starts, you name it.

The world oil production peak represents an unprecedented economic crisis that will wreak havoc on national economies, topple governments, alter national boundaries, provoke military strife, and challenge the continuation of civilized life. At peak, the human race will have generated a population that cannot survive on less than the amount of oil generated at peak—and after peak, the supply of oil will decline remorselessly. As that occurs, complex social and market systems will be stressed to the breaking point, obviating the possibility of a smooth ride down from the peak phenomenon.

The best information we have is that we will have passed the point of world peak oil production sometime between the years 2000 and 2008.[1] The date is inexact for several reasons. One is that the reported reserves (oil left in the ground) of private sector and nationalized oil companies tend to be routinely overestimated, variously to benefit the share price of stock or to gain export quota advantages in international markets, as in the case of Organization of Petroleum Exporting Countries (OPEC) members. Another reason is that the "peak" will tend to manifest in several years of oscillating market instability, a volatile period of recurring price shocks and consequential recessions dampening demand and price, presaging a terminal decline. The peak therefore will only be seen in a "rearview mirror" once the terminal decline begins. Signs of sustained market instability therefore tend to suggest the earlier onset of peak but would not be provable except in hindsight.

In other words, the peak may seem to be a kind of plateau or overhang for a few years as economic stagnation (i.e., lack of growth) curtails demand.

1. Among the authorities combining to predict the global oil production peak in this range are the Uppsala Hydrocarbon Study Group of the Association for the Study of Peak Oil (ASPO), chaired by Colin J. Campbell, retired geologist for Texaco, British Petroleum, Amoco, and Fina (also see Chapter 1 footnote 4); David L. Goodstein, professor of physics, California Institute of Technology; Matthew R. Simmons, CEO of Simmons & Company International, chief investment banking firm serving the oil industry; Albert Bartlett, professor emeritus, physics department, University of Colorado, Boulder; Jean Laherrère, retired geologist for the French oil company, Total; Kenneth S. Deffeyes, professor emeritus of geology at Princeton University; Walter Youngquist, retired professor of geology at the University of Oregon; L. F. Ivanhoe, coordinator of the M. King Hubbert Center for Petroleum Supply Studies in the Department of Petroleum Engineering at the Colorado School of Mines in Golden, Colorado; Cutler J. Cleveland, director of the Center for Energy and Environmental Studies at Boston University; David Pimentel, professor emeritus, entomology, ecology and systematics, Cornell University; and others.

During this rollover period, markets may use allocation strategies to keep their best (industrialized) customers supplied at the expense of the cash-starved "loser" nations of the world (once called "developing nations" but more likely to become "nations never to develop"). Then, slowly at first and at an accelerating rate, world oil production will decline, world economies and markets will exhibit increased instability with ever wilder oscillations from prepeak norms, and we will enter a new age of previously unimaginable austerity. These trends are irreversible.

How could such a catastrophe be so close at hand and civilized, educated people in free countries with free news media and transparent institutions be so uninformed about it? I am not one for conspiracies. While they have happened in history, conspiracies almost invariably have to be very small, and limited to tiny circles of individuals. Human beings are not very good at keeping secrets; individual self-interest is not interchangeable with group interest and the two are often in conflict, most particularly among small groups of plotters. I do not believe that the general ignorance about the coming catastrophic end of the cheap-oil era is the product of a conspiracy, either on the part of business or government or news media. Mostly it is a matter of cultural inertia, aggravated by collective delusion, nursed in the growth medium of comfort and complacency. Author Erik Davis has referred to this as the "consensus trance."[2]

When we think about it at all, most Americans seem to believe that oil is superabundant, if not limitless. We believe that the world is full of enormous amounts of as-yet-undiscovered oil fields, and that "new technologies" for drilling and extraction will perform prodigious miracles in extending the life of existing oil fields. For many of us, even people who ought to know better, the thinking stops here. The oil corporations know better but they also know that bad news is bad for business, and because there are no ready substitutes for oil they have decided to soft-pedal the news about world peak. Either that or they put a smiley-face spin on the situation. British Petroleum (BP) recast itself "Beyond Petroleum" in order to gain some points for social responsibility without really changing anything it does.

2. Erik Davis, *TechGnosis: Myth, Magic and Mysticism in the Age of Information*, New York: Three Rivers Press, 1998.

Colin Campbell, an oil geologist who has worked for many of the leading international oil companies, including BP, put it this way:

> The one word they don't like to talk about is depletion. That smells in the investment community, who are always looking for good news and the image, and it's not very easy for them to explain all these rather complicated things, nor indeed do they have any motive or responsibility to do so. It's not their job to look after the future of the world. Their directors are in the business to make money, for themselves primarily and for their shareholders when they can. So I think it's certainly true the oil companies shy away from the subject, they don't like to talk about it, and they are very obtuse about what they do or say about it. They themselves understand the situation as clearly as I do, and their actions speak a lot more than their words. If they had great faith in growing production for years to come, why did they not invest in new refineries? There are very few new refineries being built. Why do they merge? They merge because there's not room for them all. It's a contracting business. Why do they shed staff, why do they outsource people? BP aims to have 30 percent of its staff on contract. This is because it doesn't want long-term obligations to them. The North Sea is declining rapidly. They don't like to say so, but I think only four wildcats were drilled there this year [2002]. It's over! It's finished! And how can BP or Shell and the great European companies stand up and say, well, sorry, the North Sea is over? It's a kind of shock they don't wish to make. It's not evil, or there's no great conspiracy, or anything. It's just practical daily management. We live in a world of imagery and public relations and they do it fairly well, I'd say.[3]

3. Global Vision Rio +10 Interviews (*www.global-vision.org/wssd/campbell.html*), November 2002. Colin Campbell, from the UK, has worked with BP, Texaco, Fina, and Amoco. He was exploration manager for Aran Energy in Ireland, and has been a consultant to various governments at Shell and Esso. In 1998 he and Jean Laherrère were largely responsible for convincing the International Energy Agency that the world's output of conventional oil would peak in the following decade. He is the author of two books and numerous papers on oil depletion and has lectured and broadcast widely.

Corporate executives are subject to various other mentally disabling pressures. One is that they understandably tend to believe the economists in their employ, who are violently opposed to economic models that are not based on continual growth. Because the oil peak phenomenon essentially cancels out further industrial growth of the kind we are used to, its implications lie radically outside their economic paradigm. So the oil peak phenomenon has been discounted to about zero among conventional economists, who assume that "market signals" about oil supplies will inevitably trigger innovation, which, in turn, will cause new technology to materialize and enable further growth. If the market signals are not triggering innovation, then the problem must be overstated and growth under the oil regime will resume—after, say, a normal periodic downcycle. This is obvious casuistry, but casuistry can be a great comfort when a problem has no real solution.

Corporate executives fall victim to their own propaganda as much as the general public, in this case the wishful fantasies that someone will *come up with* alternative fuels in time for all the oil executives to retire with a clear conscience as well as a portfolio full of stock options. This kind of blind optimism is a holdover from the techno-miracle cavalcade of the twentieth century, combined with the mythic production exploits of American industry in World War II—all enabled by now-squandered domestic supplies of petroleum—which has fed the mentality of American exceptionalism.

The American government has had access to better information, too, but that information has only led to a political dilemma for administrations of both parties. Our investment in an oil-addicted way of life—specifically the American Dream of suburbia and all its trappings—is now so inordinately large that it is too late to salvage all the national wealth wasted on building it, or to continue that way of life more than a decade or so into the future. What's more, as we have outsourced manufacturing to other countries, the entire U.S. economy has become more and more dependent on continued misinvestment in American Dream suburbia and its accessories. No politician wants to tell voters that the American Dream has been canceled for a lack of energy resources. The U.S. economy would disintegrate. So, whichever party is in power has tended to ignore the issue or change the subject, or spin it into the realm of delusion—assisted by agencies such as the U.S. Geological Survey, which serve their masters very well by supplying often inaccurate but reassuring reports.

One president in recent times, Jimmy Carter, told the truth to the American public. He told us that our continued hyperdependence on oil was a deadly trap and that we would have to change the way we live in America. He was ridiculed and voted out of office for it. Of course Carter himself, having been trained as a nuclear technician in the Navy during the Sputnik era, when America rushed to catch up with the Russian space program (another heroic and successful exploit of technological research and development), tended to believe that a crash program in alternative and synthetic fuels would yield some miracle replacement for fossil fuels. And the hopes Carter planted in a series of 1979 speeches still affect our national psychology, though we are no closer to developing significant replacements for oil a quarter century later.

It is a little hard to say what Ronald Reagan and the first George Bush really thought about America's oil predicament, because both affected to subscribe to a branch of evangelical Protestantism that posited an "end times" apocalyptic scenario for the near future, meaning that it wouldn't matter what happened to the world very far into the twenty-first century because the kingdom of Jesus was at hand. Were Reagan and George H. W. Bush only pretending, or did they actually believe the future was irrelevant?

During the Clinton presidency, baby-boomer hippies had matured into yuppies who enjoyed the benefits of cheap oil so much (and were so spoiled by it) that they fell easily into a consensus trance regarding America's energy future: party on. The Alaskan and North Sea oil bonanzas had erased their memories of the brief 1970s oil crises. During most of the 1980s and 1990s, gas prices at the pump were lower in constant dollars than at any other time in history. It was the former-hippie boomer yuppies, after all, who started the SUV craze and bought the McMansions way off in the outermost suburbs. At the same time, stunning advances in computer development (boomer-led), and the rapid growth of the huge new industry that went with it, had induced among the boomer cultural elite a mentality of extreme techno-hubris, leading many to the conviction that our fantastic innovative skills guaranteed a smooth transition into the alternative fuels future—which, of course, squared with the wishful views of conventional economists. It all amounted to an unfortunate self-reinforcing feedback loop of delusion. Clinton Democrats regarded any upticks in oil prices as being a conspiracy between the Republicans and their donor-sponsors in the oil industry. Meanwhile, Democrats have tried to compensate for their purblind irresponsibility on energy issues by assuming a position of moral superiority on environmental issues. Yet many yuppie progressive "greens" are the ones who drove their SUVs to environmental rallies and, even worse, made their homes at the far exurban fringe, requiring massive car dependence in their daily lives. The epitome of this attitude was Amory B. Lovins, head of the Rocky Mountain Institute, who devoted his organization's time and energy in the 1990s to the development of a high-mileage "hypercar" that would have only promoted the unhelpful idea that Americans can continue to lead urban lives in the rural setting. Lovins also built the organization's headquarters in a remote part of the Colorado backcountry, which employees could get to only by car.

It can be stated with certainty that George W. Bush was fully informed of the hazards of the oil peak situation by at least one credible authority, Matthew Simmons, a leading oil industry investment banker and a highly regarded public commentator who has spoken forthrightly in scores of conferences and symposia about the hazards presented by the coming global oil peak. Simmons was brought in to advise the Bush campaign as early as 1999 and had many frank discussions with Bush both before and after the election.[4] Of course, the younger George Bush, like his father, as well as being an "oil man," was a self-professed evangelical Christian and in the background of his belief system lurked that dark idea that Armageddon was just around the corner. Could he be relied on to care about the future?

What's So Special About Fossil Fuels, Anyway?

Fossil fuels are a unique endowment of geologic history that allow human beings to artificially and temporarily extend the carrying capacity of our

4. Julian Darley interview with Matthew Simmons, February 10, 2003, *www.globalpublicmedia.com*. Text of speeches by Simmons available at *www.simmonsco-intl.com*.

habitat on the planet Earth. Before fossil fuels—namely, coal, oil, and natural gas—came into general use, fewer than one billion human beings inhabited the earth. Today, after roughly two centuries of fossil fuels, and with extraction now at an all-time high, the planet supports six and a half billion people. Subtract the fossil fuels and the human race has an obvious problem. The fossil fuel bonanza was a one-time deal, and the interval we have enjoyed it in has been an anomalous period of human history. It has lasted long enough for the people now living in the advanced industrialized nations to consider it absolutely normative. Fossil fuels provided for each person in an industrialized country the equivalent of having hundreds of slaves constantly at his or her disposal. We are now unable to imagine a life without them—or think within a different socioeconomic model—and therefore we are unprepared for what is coming.

Oil and gas were generally so cheap and plentiful throughout the twentieth century that even those in the lowest ranks of the social order enjoyed its benefits—electrified homes, cars, televisions, air conditioning. Oil is an amazing substance. It stores a tremendous amount of energy per weight and volume. It is easy to transport. It stores easily at regular air temperature in unpressurized metal tanks, and it can sit there indefinitely without degrading. You can pump it through a pipe, you can send it all over the world in ships, you can haul it around in trains, cars, and trucks, you can even fly it in tanker planes and refuel other airplanes in flight. It is flammable but has proven to be safe to handle with a modest amount of care by people with double-digit IQs. It can be refined by straightforward distillation into many grades of fuel—gasoline, diesel, kerosene, aviation fuel, heating oil—and into innumerable useful products—plastics, paints, pharmaceuticals, fabrics, lubricants.

Nothing really matches oil for power, versatility, transportability, or ease of storage. It is all these things, plus it has been cheap and plentiful. As we shall see later, the lack of these qualities is among the problems with the putative alternative fuels proposed for the post-cheap-energy era. Cheap, abundant, versatile. Oil led the human race to a threshold of nearly godlike power to transform the world. It was right there in the ground, easy to get. We used it as if there was no tomorrow. Now there may not be one. That's how special oil has been.

Where Oil Comes From

Oil is ancient organic matter that has been heated under tremendous pressure and transformed chemically into chains and clusters of hydrogen and carbon atoms. Hydrogen and carbon are chemical elements, substances that cannot be reduced further in terms of atomic structure. The chains and clusters of atoms are called molecules. These molecules are the basic building blocks of compounds, in this case hydrocarbons. The lightest hydrocarbons, such as methane and propane gases, are made of molecules containing very few hydrogen and carbon atoms. Liquid hydrocarbons, such as gasoline and lubricating oils, contain more atoms per molecule. The very heavy hydrocarbons, such as tar and paraffin wax, contain more complex clusters and chains of hydrogen and carbon atoms and come in the form of semisolids and solids.

The organic matter that started the oil forming process was algae thought to have "bloomed" in the shallow reaches of prehistoric lakes and oceans during long, favorable periods of prehistoric global warming, between 300 million and 30 million years ago. This accumulated goop of dead plant matter, called kerogen, built up into underwater sediments, which were later thrust down or folded by movements in the earth's crust. Eventually tectonic forces subducted them to a depth between 7,500 and 15,000 feet. Among the many things geologists have learned during the bloom in science that accompanied the industrial era is that temperatures under the earth increase by about 14 degrees (Fahrenheit) per 1,000 feet. Temperatures (and high pressure) at depths between 7,500 and 15,000 are just right for cooking ancient kerogen-containing sediments into hydrocarbon saturated sedimentary rock. At depths below 15,000 feet, pressures are so great and temperatures are so high that all the hydrocarbon molecules break down into the simplest hydrocarbon compound, methane gas, made of one carbon atom hooked to four hydrogen atoms, much of which escapes through layers of rock over time. Therefore, the depth between 7,500 and 15,000 feet is called the "oil window." Outside of this window, oil is not likely to be formed.[5]

5. Kenneth S. Deffeyes, *Hubbert's Peak, The Impending World Oil Shortage*, Princeton, NJ: Princeton University Press, 2001.

It is obvious that while oil may form in this underground window, it is often discovered in layers above 7,500 feet. Oil sometimes even works its way up to the surface. Petroleum seeps are instances where underground pressure creates oil pools above ground. Seepages occurred in ancient times all around the oil-rich Middle East. Humans have long known about this peculiar, semimagical substance. The peoples of the biblical and classical world used tar to caulk boats and to pave streets, and flung flaming clots of the stuff at each other in naval warfare. Asphalt was used as a mortar in both Jericho and Babylon and was traded as a commodity around the Mediterranean. Oil has long been used as a medical nostrum for every sort of illness and complaint.

A crude kind of petroleum mining industry arose in the early nineteenth century around surface seeps in what is now Romania, and small amounts of kerosene were distilled from it, though the development of a clean-burning lamp lagged behind. Surface petroleum was called "rock oil" in America back before the industry got under way in earnest.[6] The famous LaBrea tar pits of Los Angeles are an example of seepage (not incidentally, in a region of high tectonic activity).

What accounts for the presence of oil far above the window is continuous tectonic movement, upthrust from below, plus erosion working away the top layers over many scores of millions of years. So, petroleum can "work its way" to the surface in some rare places. The Athabaska tar sands of Canada are ancient oil deposits that became exposed by geologic action, allowing the lighter petroleum liquids to evaporate off over eons of time.

Much of the oil discovered and extracted during the mature years of the oil industry, however, has been found at precisely those depths within the oil window because continental masses pass only intermittently over tectonic "hot spots" where violent geologic uplift and downthrust occur. "The practical result for the oil industry," Kenneth Deffeyes writes, "is that sediments on top of the stable continents dance in and out of the oil window for geologically long times."

Oil formed in the way I have described is concentrated in discrete "basins" or "fields" in particular parts of the world and not in others. These

6. Daniel Yergin, *The Prize: The Epic Quest for Oil, Money and Power*, Touchstone, 1993.

are relatively small, special areas relative to the total geologic surface of the earth, and they are deployed in such a way as to suggest that the original material was laid down along ancient sea basins in places where nutrients were especially well "trapped" for the nourishment of kerogen-forming organisms—perhaps where ancient rivers entered bays or where rainfall favorably affected the salinity of the water and growth of algae.

What is known, therefore, about the geology of oil suggests that there are no inexhaustible reserves below the window, and that it is very unlikely that known oil fields or basins can be "replenished" from some mysterious source far deeper beneath the earth's crust, as some wishful commentators would like to believe. In fact, all credible authorities agree that these special and precious pockets of fossil hydrocarbons will all be gone by the end of the twenty-first century even if the rate at which we use them doesn't increase. The complacency over this rather startling fact is due to the presumption that technology and markets will naturally rescue us. But we will have tremendous problems over oil as soon as it has passed the arc over the global peak and we commence the slide down the other side of the curve.

Oil and Industry

The modern oil industry got under way in August 1859, when a colorful figure named Edwin L. Drake drilled into the ground near a surface seep on a farm in northwestern Pennsylvania, using steam-driven technology designed for water wells, and struck oil under pressure at seventy-five feet. It had been thought that oil was a kind of liquid residue of coal deposits, with which Pennsylvania was amply endowed. The oil didn't gush out of Drake's first well as it does in movies, but enough flowed up the pipe to fill every empty barrel in the vicinity. Drake's had been an experimental venture, backed by a handful of wealthy sponsors back in New Haven, Connecticut, who saw potential value in the substance if it could be gotten in greater volume than the few bucketfuls a day that seeps typically produced. Nobody had ever tried drilling for the stuff before Drake.

The rise of the oil industry in America really must be viewed against the greater backdrop of two other parallel historical narratives: the continuing worldwide industrial revolution and the continued settlement of the American continent, including the exploitation of its extensive resources, among which would prove to be a lot of oil, which in turn would end up powering the most robust phase of the industrial revolution.

When Drake drilled that well, the industrial age was already well established, powered by wood, coal, and water. The initial great promise of oil was for illumination. The population was growing and people were moving from farms to burgeoning industrial towns and cities where the need for indoor lighting was increasing tremendously. The highest-quality illuminant then known was sperm whale oil. It was hard to get, expensive, and the limited supply was diminishing as whales were hunted remorselessly. Meanwhile, in many towns and cities gasworks had been established that cooked coal to distill off illuminating gas, but its main application was for street lighting. In the home, gaslights had several drawbacks. They were noisy, hot, dangerous, and couldn't be moved around the room because the lamps had to be fixed to their pipes on the wall or ceiling. However, the new distillate of rock oil, kerosene, was comparable to whale oil in brilliance. It was not explosive the way gas was, it didn't make noise as it burned, and the lamps could be moved anywhere in a room as required. Unlike gaslight, kerosene required no costly infrastructure of pipes for delivery to the point-of-use. Kerosene lighting took off sensationally.

In the new factories springing up everywhere, faster-moving, hotter-running machinery required new oil-based lubricants to replace old standbys such as lard. Many other uses for petroleum products would follow. After Drake's strike, on the eve of the Civil War, a boom in oil not unlike a gold rush occurred. Drilling derricks popped up all over western Pennsylvania. Just as suddenly, a glut developed. The business took on the boom-and-bust character that would define much of its future. The amount of oil coming out of the ground soon far exceeded the ability of the system to allocate it. Producers had to use the same expensive wooden barrels made for whisky and pickles. The railroad tank car hadn't been invented yet, nor had the overland pipeline. Refineries were little more than backyard science projects. Even the manufacture of household lamps to burn kerosene lagged. For a little historical perspective, consider this: During the Civil War, kerosene lamps were new technology.

I won't recapitulate the whole history of the oil industry—which has been done so well by Daniel Yergin, among others. In summary, beginning in the late 1860s and over the next thirty years, John D. Rockefeller managed to consolidate most of the refining and marketing capability of the new oil industry until he had leveraged his way with pricing power to a near-imperial monopoly, Standard Oil. By this time, the great European nations had gotten into the act in the Old World. By the 1880s, the Rothschilds; Alfred Nobel's lesser-known brother, Ludwig; and Englishman Marcus Samuel all formed companies to exploit the oil region around Baku in southern Russia near the Caspian Sea. In the 1890s, the Royal Dutch Company opened up what is now Indonesia to exploration and drilling. They all battled with Standard Oil for world markets, and several of them, such as Samuel's Shell and Royal Dutch, would go on to become global giants.

Back in America, the oil fields of the East systematically played out, and the business of extraction moved westward to Ohio and Illinois and finally to Texas and Oklahoma, where gigantic "elephant" oil fields were discovered—among them the 1901 Spindletop "gusher" near Beaumont. In 1911, the U.S. government broke up the Standard Oil monopoly, though each deconstructed piece eventually became a major oil company in its own right. Around the same time, the focus of the oil business shifted from illumination and lubricants to gasoline for the recently invented automobile. Electrification in the cities was quickly replacing kerosene lighting, which more and more became an emblem of rural backwardness. Automobile sales exploded after Henry Ford devised his assembly-line system of production in 1913, and the price of his Model T fell steadily year after year afterward. As motoring became democratized, a furious nationwide project of road-building and paving created a huge demand for the sludgy by-products of gasoline refining: asphalt and tar.

Between 1880 and 1930, the major cities of America grew to a scale that had never been seen before in world history. New York, Chicago, and Detroit came to represent a futuristic urbanism of immense megastructures, including both office towers such as the Chrysler Building and vast horizontal factories like Ford's River Rouge plant. (The skyscraper would not appear in Europe until the 1960s, and even later in Asia.) It was assumed by intellectuals, especially in the fields of architecture and urbanism, that human life had crossed an evolutionary threshold that rendered previous human history and tradition obsolete. The new industrial man rushed upward to seemingly godlike powers on wings of gasoline-powered technology. Modernism became a kind of secular religion. Machine aesthetics reigned. In politics, the Modernist machine ethos found an analogue in fascism, Nazism, and Soviet communism.

The practice of war, too, switched into an oil-powered industrial mode. The nineteenth century after Waterloo had been relatively peaceful in Europe except for the fiasco in the Crimea (1853–56), fought in a geographical backwater between England, France, Russia, and Turkey, and the Franco-Prussian War (1870–71), which lasted barely six months, half of which consisted of the static siege of Paris. In the absence of major conflict, tactics and strategy did not catch up with technology. Thus, in the portentous summer of 1914, the armies of Britain, France, and Germany marched gaily off behind their mounted officers, their cannon hauled by horses, as if going to a dramatic reenactment of past wars. They expected quick and decisive victories, only to find themselves bogged down by machine-gun fire in muddy trenches for the next four years. The scale of the slaughter in France and Belgium was analogous to the gigantic new scale of industry and cities in the developed nations. It made even the American Civil War seem like a gang rumble in comparison. In the first Battle of the Marne alone (September 1914), French and German casualties each totaled about 250,000.

The combatants motorized rapidly, desperately hoping to gain some advantage by using the power of oil. Yergin writes that the British army took only 827 motorcars and a mere fifteen motorcycles off to France in 1914. Four years later, they had accumulated 56,000 trucks, 23,000 cars, and 34,000 motorcycles—a startling feat of production in its own right. When the Americans joined the conflict in April 1917, they brought another 50,000 motor vehicles with them.[7]

The farsighted Winston Churchill, then Lord of the Admiralty, converted all of England's warships from coal to oil in 1912 just in time for the outbreak of hostilities. The conversion significantly increased the ships' range

7. Yergin, p. 171.

and speed compared with the coal-powered German fleet, and afforded the allies an effective blockade against German shipping. To ensure the fuel supply, the British government bought a majority share position in the Anglo-Persian Oil Company that had been set up to exploit newly discovered fields in what today is Iran, with transport through the Suez Canal. The Germans, on the other hand, rapidly deployed new diesel-powered submarines to undercut British superiority in surface craft. The U-boat campaign was at first remarkably successful in choking off supplies to England and France, but eventually tactics caught up with technology and the allies learned the art of the naval convoy. However, Germany's aggressive U-boat campaign had the negative effect of drawing the United States into the conflict against the Germans.

Germany suffered, too, from having poor access to oil supplies. Their main source was Romania, which took sides in 1916 with Russia, Germany's enemy. When the Germans then attempted to capture the Romanian oil fields, a British commando force got there first via the Black Sea and blew them up. Petroleum starvation, along with the 1918 flu epidemic, eventually put the German war machine out of business, while the British innovation of the tank also helped break the stalemate of the trenches. World War I also saw the first use of airplanes in combat, though their use was a limited and romantic sideshow to the action on the ground.

World War II was fought with oil and for oil, as both Germany and Japan tried desperately to extend their hegemony to distant oil-producing regions to ensure the continuation of their rapidly developed industrial economies. Both lost the war, in large part, because they failed to secure the oil. The Japanese struck south to establish control over Indonesia, where proven oil fields had been in production since the 1890s. Among the U.S. Navy's most successful tactics was the systematic sinking of Japanese oil tankers until the enemy military machine was starved for fuel. Germany ventured into southern Russia to secure the Baku region (and to knock Russia out of the war by depriving it of its main oil supply). But the Germans got bogged down along the way at Stalingrad in the winter of 1943, retreated in disarray, and thereafter had to rely on synthetic liquid fuels manufactured at home out of coal.

The World Leader

Among the nations that had started down the path of industrialization in the nineteenth century, America was by far the best endowed with oil. Discovery of new fields around the United States increased tremendously in the 1920s and 1930s as geologists explored the Far West and California. In fact, U.S. discovery peaked in the 1930s, but it would be decades before this or its implications were understood. Oil was so plentiful in the United States during the Great Depression, and demand so slack due to the crisis in finance, that prices slumped to ten cents a barrel. At the outbreak of World War II, America was figuratively drowning in oil. There was no need to import it from faraway places by tanker, as the Europeans did.

The United States was the first industrial nation to find and exploit oil in commercial quantities, and the first to make widespread application of oil in primary manufacturing, transportation, and consumer products—indeed, to elaborate the whole system of a modern consumer economy. The automobile industry and all its interrelated enterprises started in the United States. The plastics and synthetic fibers industries were born here, using oil as the feedstock. During the hundred-plus-year run of the world oil economy, the United States has never ceased to be the world's leading consumer of oil.

The United States remained the world's leading oil producer and exporter for much of the twentieth century. The stupendous destruction and disruptions attending World War II left America preeminent in the oil business for a quarter-century afterward. Europe and Asia were ravaged by the war, but the American mainland was not touched. The United States emerged with its manufacturing and oil infrastructure intact. Though discovery of oil—that is, the finding of new fields—had fallen off by then, the rate of production from fields already found would continue to rise until 1970. Where oil geology is concerned, North America may be the most minutely explored continent on earth. There are virtually no unknown oil fields of any significance left to be found here. No amount or kind of new technology will alter that.

During the 1950s America's chief interest was in reestablishing political and economic stability in the world after the debacle of two world wars and a depression. An ethos of genuine beneficence stood behind this policy, a wish born out of victory and self-sufficiency to do good in the world after a decade of extreme violence. It included the discipline to refrain from punishing former enemies, a determination to act humanistically as a counter to the appalling inhumanity of the Nazis, and a definite nervousness about the atomic future.

France and England were financially exhausted, while Germany and Japan lay in ruins, and exploiting them economically would not have been possible. The Americans' new adversary, Russia, had plenty of its own oil. Arabia was just coming into production in a big way. For a while, the world had quite a bit more oil than it could use. The 1950s did become a period of extraordinary equilibrium, prosperity, and hopefulness at home, and the experience of growing up in that optimistic decade forms the baseline of expectations for many baby boomers who today run the nation.

After World War II, the American public made two momentous and related decisions. First was the decision to resume the project of suburbanization begun in the 1920s and halted by the Great Depression and war. By the 1950s, the prevailing image of city life was Ralph Kramden's squalid tenement apartment on television's *The Honeymooners* show. Suburbia was the prescribed antidote to the dreariness of the hypertrophied industrial city—and most American cities had never been anything but that. They were short on amenity, overcrowded, and artless. Americans were sick of them and saw no way to improve them. Historically, a powerful sentimental bias for country life ruled the national imagination. As late as 1900, the majority of U.S. citizens had lived on farms and American culture was still imbued with rural values. As far as many Americans were concerned in the 1950s, suburbia *was* country living. There was plenty of cheap, open rural land to build on outside the cities, and as soon as mass-production house builders like William Levitt demonstrated how it might be done, suburbia would be thoroughly democratized—country living for everyone. That suburbia turned out to be a disappointing cartoon of country living rather than the real thing was a tragic unanticipated consequence, which I have described in my previous books.[8]

8. *The Geography of Nowhere*, New York: Simon and Schuster, 1993; *Home from Nowhere*, New York: Simon and Schuster, 1996; *The City in Mind: Notes on the Urban Condition*, New York: Free Press / Simon and Schuster, 2002.

Second was the political decision in 1955 to build the interstate highway system, the largest public works project in the history of the world. It has been argued elsewhere that the interstate system was conceived for moving troops and evacuating cities in the event of a future war; also that it was an economic stimulus program to prevent a return of the dreaded prewar depression. Perhaps, but I would argue that the public was simply entranced by cars and wanted a state-of-the-art nationwide road system as a kind of present to ourselves for winning the war. It was not anticipated at the time that the interstates would lead to the catastrophic disinvestment in U.S. cities. Nor did many foresee the debasement of the rural landscape as the office parks, strip malls, chain stores, and fry-pits eventually settled beside the freeway off-ramps and came to occupy every hill and dale between the housing developments. The consensus in the 1950s was that suburbanites would continue to commute into the city for work, shopping, and entertainment.

In any case, the result of these two decisions was technological lock-in. Once the investment was made in the infrastructure and furnishings of suburbia, we were stuck with it, and with the enormous amounts of oil required to run it.

Hubbert's Curve I—American Peak

It was inherent to the peculiar boom-and-bust nature of the oil industry that from the very start much uncertainty and anxiety existed about how much oil was actually hidden below the earth's surface. Was it common or rare? Was it all over or just in a few special places? Might there be a limitless reservoir of the stuff deep inside the earth? No one really knew. The science developed only with the industry itself—the understanding of what kinds of subterranean structures might contain oil. A clear picture of where oil was distributed around the nation, not to mention the globe, had yet to resolve. In the early decades, the Pennsylvania fields had played out rather quickly using primitive methods of extraction, and the fields of Ohio and Indiana were drained rapidly after that. Car ownership had exploded from 1.8 million to 9.2 million between 1914 and 1920. In 1919, the U.S. Bureau of Mines predicted that U.S. oil production would enter terminal decline in the mid-1920s. Some thought the petroleum craze would prove as transient as the sperm whale oil era. Then the monster fields of East Texas and Oklahoma were discovered and California followed. Elsewhere, giant fields were found in Mexico, Venezuela, Persia (now Iran), the East Indies, and Central Asia. The great discoveries in Arabia lay decades ahead.

After 1945, America's position in the world vis-à-vis oil was special and privileged, perhaps to a degree that remains less than fully appreciated. Europe, having fought over distant supplies of oil in two world wars and suffered hugely, never became complacent about it, as reflected in Europeans' compact living arrangements and their high luxury taxes on gasoline. But America, having won those wars and possessing substantial reserves of oil in situ, became overconfident to a dangerous degree about its oil future. When a geologist named M. King Hubbert announced in 1949 that there was, in fact, a set geological limit to the supply of oil that could be described mathematically, and that it didn't lie that far off in the future, nobody wanted to believe him. Hubbert was not a lightweight. Before World War II, he had taught geology at Columbia University and worked for the United States Geological Survey. His theoretical work on the behavior of rock in the earth's crust was highly regarded and led to innovations in oil exploration. Stretching from 1903 to 1989, Hubbert's whole life took place during the high tide of the oil era, and he played a large part in developing its science. But he was a visionary who dared to imagine the final act of the oil drama.

By the mid-1950s, as chief of research for Shell Oil, Hubbert had worked up a series of mathematical models based on known U.S. oil reserves, typical rates of production, and apparent rates of consumption, and, in 1956, he concluded that the oil production in the United States would peak sometime between 1966 and 1972. Hubbert also demonstrated that the rate of discovery would plot out a parallel path as the rate of production, only decades earlier. Since discovery of new fields in the United States had peaked in the 1930s and declined remorselessly afterward, despite greatly improved techniques in exploration, the conclusion was obvious. Production declines would follow inexorably, Hubbert predicted, despite improved drilling and extraction methods. After this point of maximum production, or "peak," U.S. oil fields would enter a steady and irreversible arc of depletion. He displayed this information in a simple bell curve. The peak was the top of the curve. Nobody took "Hubbert's curve" seriously. His was a lone voice in a nation that was having too much fun cruising for burgers to imagine what really lay down the road.

The extraordinary rate of discovery in other parts of the world during the 1950s and 1960s reinforced American complacency, because it tended to suggest that more oil could always be found elsewhere, especially in third-world places where compliant peoples would be happy to benefit from its development. Proven world reserves, Yergin writes, had increased from 62 billion barrels in 1948 to 534 billion in 1972, almost all of it outside the United States (and the communist nations), and more than 80 percent of it in the Middle East.[9] A twenty-year glut of oil developed. The Soviets were giving it away literally at half market price because it was one of the few things they could sell to get hard currency. A small amount of fungible foreign crude found its way into U.S. markets, but for most of the 1950s and 1960s a complex system of import quotas kept most foreign oil out of the U.S. market, while domestic production operated well below capacity. Because the global oil industry was dominated at the time so overwhelmingly by American companies, their market allocation systems, and their technical expertise, there was an unrealistic assumption by Americans that these favorable conditions would continue indefinitely. This sense of invulnerability was reinforced when, following the stunning Israeli victory over Egypt in the Six-Day War of 1967, Saudi Arabia attempted to orchestrate an embargo against nations that supported Israel. America's surplus production capacity, its ability to just pump more oil as needed, allowed the West to work around Arab sanctions. The embargo fizzled. Within a few years, though, everything changed.

U.S. oil production proceeded to peak in 1970—though the peak was not detected until the following year, when lower figures started to come in. Peak production in 1970 was 11.3 million barrels a day. That would be the highest level ever, and production would fall by several percentage points a year ever after. (By the mid-1980s total U.S. crude production fell under 9 million barrels a day and is currently under 6 million.) Meanwhile, in 1970, aggregate U.S. demand would cross the

9. Yergin, p. 500.

line of total U.S. production. Surplus capacity was gone. Hubbert's prediction had been absolutely correct. Two more years of denial, confusion, and inaction followed (aggravated by the national preoccupation with the growing fiasco in Vietnam and the Watergate scandal), and then the United States received a very upsetting wake-up call: the OPEC oil embargo of 1973.

The crisis occurred for a very simple reason: The United States had lost pricing power over globally traded oil because, having passed peak, it was pumping its oil at the maximum rate. What's more, as the United States passed peak, net imports rose swiftly from 2.2 million barrels just before peak to 6 million in 1973. Suddenly the United States was importing roughly a third of its oil. Without surplus capacity, the ability to open the valves and flood the market with "product," the United States had ceded control of world oil prices to somebody else who still did have surplus capacity. That "somebody else" was the Organization of Petroleum Exporting Countries (OPEC), led by Saudi Arabia.

The First Real Oil Shock

In 1973, Saudi Arabia had tremendous surplus capacity. The country had more oil to start with than the United States had had a hundred years earlier, and had entered the process of discovery and production much later than America did. The first development concessions were negotiated in the 1930s, exploration barely got under way when World War II interrupted it, and production didn't ramp up until after the war was over. By the 1970s, as America passed its all-time production peak, Saudi Arabian oil production was just entering its robust phase. Production was run by a consortium called Aramco, a joint venture among Exxon, Texaco, Mobil, and Standard Oil of California (Socal, later Chevron).

In the early 1970s, oil demand worldwide was surging, up from 19 million barrels per day in 1960 to 44 million a day in 1972. Europe had finally recovered from the war, and the backwaters of the world were beginning to "develop"—i.e., to build factories, power plants, roads, and to run automobiles. The twenty-year-long postwar world oil glut was over. Demand was ratcheting up to supply.

Meanwhile, politics in the Middle East had evolved dangerously. The Soviet Union was egging on two of its client states, Egypt and Syria, to wage another war on Israel, in the interest of pitting Israel's allies, the United States in particular, against the oil-rich Islamic nations on whom now even the United States was dependent for the most indispensable commodity of industrial capitalism. The desert oil kingdoms in turn were maturing, enjoying high birth rates, growing tremendously wealthy, modernizing, and chafing under U.S. corporate domination. The sons of sheikhs and princes were returning home from America with diplomas from Harvard and UCLA and expanded worldviews.

Aramco had been accustomed to setting the price for the oil produced in Saudi Arabia—while paying a hefty premium to the kingdom on each barrel sold—an arrantly colonialist practice, increasingly resented. Through the 1960s, the Saudis had moved incrementally from having no say in pricing, to partial say, to now wanting sovereign decision-making power over it. As the United States passed peak production, the Saudis caught on to the fact that the exporting nations (namely themselves) were now the world's swing producers, and OPEC's new mission was to organize that pricing power. Aramco and the other American "majors" operating around the Persian Gulf had little choice but to go along with whatever the Arabs decided to do, knowing that their vast investments in equipment could be seized and nationalized at any time just as Egypt had done with the Suez Canal in 1956.

The 1973 Yom Kippur war was the precipitating incident of the OPEC embargo. On October 6, Egyptian and Syrian forces caught the Israeli military off-guard on the most solemn Jewish holiday, when many soldiers were home with their families. Because the Arab-Israeli dispute was commonly viewed as yet another cold war proxy battle, the United States and its allies naturally lined up behind Israel against the Soviet-sponsored aggressors. Egypt's President Anwar Sadat implored the Saudis and other Muslim states to use the "oil weapon" against Israel's allies. On October 12, the Saudi-led OPEC demanded of the various Western companies doing business in the Middle East, including Aramco, a 100 percent increase in the posted price of their cartel's oil. The companies stalled for time. On October 16, the Persian Gulf region OPEC members broke off negotiations with the Western oil companies and announced that thereafter they would set prices themselves. On October 17, the Israelis gained the upper hand on the battlefield, thanks in large part to aggressive American resupply efforts, and began to push the Egyptians back across the Sinai peninsula and the Syrians out of the Golan Heights. The same day, the Arab oil ministers announced an oil embargo on the United States, while increasing prices by 70 percent to western Europe. Overnight, the price of a barrel of oil to these nations rose from $3 to $5.11. On October 19, President Richard M. Nixon announced a military aid package for Israel. The following day, Saudi Arabia retaliated by announcing a total cutoff of oil exports to America.

Effects of the Embargo

A UN ceasefire ended hostilities on October 22, 1973, but the OPEC embargo against the United States remained in force while the organization further increased the price per barrel to the rest of the world. What followed was an interesting case study in network breakdown and cascading failure. In fact, the embargo never actually achieved a shutoff of OPEC oil imports to the United States. All but about 5 percent of the needed supply found its way to America by a circuitous route as allocations to other nations were surreptitiously redirected. But the base price of a barrel of oil did eventually more than quadruple by the time the embargo was called off in March 1974. And the price rise alone staggered the West and Japan.

Already at that time, public transit was a thing of the past and about 85 percent of Americans drove to work every day. Now lines of cars formed at the gas stations; posted prices changed hourly in some places as station owners took advantage of a panic situation; fights broke out among the motorists waiting on line. Odd and even license plate numbers were used to assign gas-buying privileges on certain days of the week. The industry's own national allocation system failed and some parts of the United States got plenty of gasoline while others got none at all, which reinforced the confusion and panic. In November, President Nixon, otherwise consumed with Watergate, proposed an extension of daylight saving time and a total ban on the sale of gasoline on Sundays. Both were later approved by Congress. Ration stamps were printed but never issued.

The U.S. economy suffered a body blow. Because absolutely everything in the industrial economy was either made, transported, or carried on with petroleum products, the price rise alone thundered through the system. Prices of food and manufactured goods shot up. The entire American workforce suffered, in effect, a substantial cut in pay. The stock market dropped 15 percent in a month, and a year later it would be down 45 percent from its pre-embargo high. The U.S. auto industry, the crown jewel of the economy, was devastated. The "Big Three" were all tooled up to produce fleets of oversized, entropic, gas-guzzling behemoths, while the Europeans and now the Japanese offered small, nimble, fuel efficient, and better-built models. America's compact cars were a long-standing joke, in particular General Motors Chevrolet Corvair, castigated by consumer advocate Ralph Nader as *Unsafe at Any Speed*. The 1973 oil crisis made American cars look ridiculous even to Americans, and plummeting sales soon reflected this starkly. The Big Three never recovered their market preeminence.

Price inflation—with annual rates as high as 12.8 percent in the United States and even higher in Europe—led to monetary and fiscal distortions as interest rates increased to scale. Institutions lending money had to protect themselves. Under such inflationary conditions creditors would be repaid in dollars worth less than the dollars lent out, so loans and mortgages carried inflated premiums. Such extremely high interest rates on business loans made it difficult for companies to rationally allocate resources for capital expenditure. High interest rates discouraged house buying because the mortgage interest was gargantuan. The industrial nations entered deep recessions, the worst since the 1930s. The so-called developing nations were especially hard-hit. Many of those in Africa that had only recently emerged from colonialism in the 1960s would be permanently saddled with debt after the OPEC embargo and would never make a successful transition into sovereign self-sufficiency.

By the time Richard Nixon was finally forced out of office by the threat of impeachment and the likelihood of conviction, America had entered a peculiar zone of economic horse latitudes dubbed "stagflation" (stagnation + inflation). Many mainstream economists affected to be mystified by the phenomenon because in their models and paradigms the phenomenon of inflation was a handmaiden of vigorous economic expansion.

What many of them failed to register was the way that the OPEC embargo and its aftershocks represented a unique crisis of a fossil fuel-addicted industrial civilization—the suspension of assumed growth per se.

Though the embargo officially ended in March 1974, the high prices per barrel remained in effect and the economic effects lingered for years. Saudi Arabia bought out the U.S. companies' interests in Aramco and nationalized its oil fields in 1975. From then on, Aramco ran the Arabian oil fields as a mere hired manager.

President Jimmy Carter attempted to awaken the American public to the idea that the energy crisis was a more or less permanent condition reflecting the real drawdown of the nation's number one nonrenewable resource. Carter attempted to fashion a coherent national energy policy; passed tax and rate incentives for hydroelectric development, especially on the small, local scale; restarted Nixon's "Project Independence" to develop synthetic hydrocarbon and alternative fuels; and set the tone at the top in a rhetoric that refused to soft-soap the problem. Appearing on television three months after his inauguration in April 1977, wearing a cardigan sweater and seated by a blazing fireplace, Carter declared that the nation's energy predicament was the "moral equivalent of war." He even installed passive solar water heaters on the White House roof. Carter's legitimacy would soon be fatally compromised by the standoff over American hostages following the overthrow of the shah in Iran, a year-long melodrama that distracted, preoccupied, and finally bamboozled the American public.

Hubbert's Curve II—The Worldwide Peak

The OPEC embargo brought home the frightening implications of the U.S. oil peak. But M. King Hubbert continued to research the oil depletion story he had pioneered. He eventually came up with a new model that expressed the coming global production peak, the point at which the highest rate of global annual oil production would occur, with a subsequent steady rate of annual falloff thereafter along the depletion. The model was fairly straightforward: Compare consumption with known reserves, then make some modest educated guesses about future rates of consumption. The only tricky part was knowing what was actually in the ground, because all entities engaged in oil production, both private companies and nations, tended to be cagey about what they had, for reasons already discussed. But within the drilling-and-exploration community news of major strikes always got out. Most of the world's oil, in fact, came from a score or so of giant oil fields, called "elephants." By the 1980s, the world had been geologically mapped to the extent that elephant oil fields on the scale of East Texas or the Ghawar field of Saudi Arabia were unlikely to have evaded discovery. The Russians, unhampered by conventional business constraints, had been especially avid in exploring their vast Siberian territories. World oil reserves were pretty much all accounted for. Hubbert lived a long time, and by the 1980s, the "rearview mirror effect" showed that world discovery had indeed peaked in the 1960s. Most importantly, the decreasing rate of discoveries comprised only small fields of minor consequence, which played out quickly. Hubbert's previous estimate about America's peak had been based on his theory, proved correct, that peak discovery preceded peak production by roughly thirty years. Hubbert initially estimated that the world peak would occur between 1990 and 2000. He was a little off, but not by much. Some experts think the world had, in fact, entered the "bumpy plateau" of the global production peak in the early 2000s, but it was a little too early to get a clear view via the rearview mirror effect.

Subsequent tweaking of Hubbert's model by Kenneth Deffeyes of Princeton; Colin J. Campbell, retired chief of research for Shell Oil; Albert Bartlett of the University of Colorado; and others following Hubbert's 1989 death put the peak somewhere between 2000 and 2010. At the time of this writing, Campbell had estimated the peak coming in 2007 and Deffeyes at 2005. Where world economic and political affairs are concerned, the difference is inconsequential.

The earth's total endowment of conventional liquid petroleum was estimated to be roughly two trillion (2,000 billion) barrels. At peak we will have extracted and burned half of that endowment. Virtually all of the consumption to date took place after 1859 and the bulk of it was disproportionately consumed in only the past fifty years, so the entire oil age, from birth to high point, has been historically very brief. According to Hubbert's model, and assuming at least current levels of world petroleum consumption at 27 billion barrels a year (and notwithstanding a still hugely expanding world population and the continued rapid industrialization of China), then the world has only about thirty-seven years of oil left, in the ideal case that every last drop is pumped out. Of course, it is extremely unlikely that the human race will ever completely drain the world's oil fields. Long before a field gives up its final barrels, the oil becomes so difficult or expensive to pump out of the ground that it takes more than a barrel of oil's worth of energy to pump out one barrel. (Many wells are currently taken out of production when they are only half-depleted because of the economics of extraction.) In the meantime, there will be degrees of difficulty and relative cost between peak and the last drop; these costs will be both attended and greatly aggravated by political competition over the remaining supplies.

An Illusory Reprieve

In the aftermath of the OPEC oil embargo, Congress approved the construction of an 800-mile pipeline—much of it over delicate tundra terrain—to transport oil from fields at the extreme northern rim of Alaska south to terminals on the Pacific Ocean. The Alaska North Slope fields had been discovered in the 1960s—in fact, they represented the last major discoveries within U.S. sovereign territory. But extremely harsh conditions above the Arctic Circle had weighed against their exploitation until the shock of the OPEC embargo. The pipeline would eventually cost $10 billion (in 1970s dollars), an investment proportionate to America's desperation over oil. The oil began to flow out of the Prudhoe Bay fields in 1978, and soon came to represent 25 percent of America's total production. The Alaska North Slope fields peaked in 1988 at 2.02 million barrels a day and declined steadily afterward. (In 2003, they produced just under one million barrels a day.)

Exploration of major oil fields in the North Sea began in earnest in the late 1960s, and discovery hit stride just before the OPEC embargo of 1973. The fields lay between the United Kingdom (Scotland, actually) and Norway, with somewhat more than half on the UK side of a mineral rights borderline that had been negotiated in 1965. Gas wells had been drilled prior to that in shallower coastal waters, but until then the offshore technology was still quite primitive, while the North Sea weather was horrendous. A new generation of deep-sea drilling platforms came along just at the right time—not incidentally, as declining discovery on the U.S. mainland led American oil companies to search deeper waters around the Gulf of Mexico and engineer new equipment to make that possible. Exploitation of the North Sea fields was extremely difficult and costly, but became imperative after the OPEC embargo of 1973. By then, the U.S. peak was incontrovertibly evident. The oil began to flow commercially in mid-1975.

Alaska and the North Sea were the last great strikes of the oil age. They allowed the West to postpone its reckoning over finite oil reserves by at least a decade. They were very productive fields while they were on the upswing, and they were controlled entirely by the West. They put enough oil onto the global markets so that supplies could not be embargoed by one bloc of nations, as OPEC had done—but price was another matter, as we shall see presently.

Great Britain benefited enormously from the North Sea bonanza. The sclerotic old empire had been suffering a wasting disease of deindustrialization since World War II and the North Sea acted like a miraculous tonic. Scotland, especially, perked up, having lost its mainstay shipbuilding industry and become a welfare basketcase just before the oil boom. The economic revival of Great Britain was thus more a manifestation of the North Sea oil boom than of Thatcherism. Norway, a much smaller nation, enjoyed even greater prosperity per capita than Great Britain from its share of the North Sea boom, but did so very quietly.

In the years just prior to the global peak, the North Sea fields would produce 9 percent of the world's total annual oil output. By the late 1990s, many of the major North Sea fields were in decline. Newer, lesser fields were playing out in as few as two years, barely enough to justify the construction of elaborate drilling platforms in the frigid, storm-tossed waters. A consensus held that the North Sea fields would peak somewhere between 2002 and 2004, though again it was somewhat early for the rearview mirror effect.[10] Since they had come online so late, and were worked with the latest drilling technology, the North Sea oil fields also tended to follow a much steeper depletion arc.

10. International Energy Outlook of the Energy Information Administration 2002 forecast. According to the report, North Sea production was likely to peak in 2004 and then gradually decline with the maturing of some of its larger and older fields. Other studies show that the North Sea already peaked in 2000 at 6.4 million barrels a day.

The Second Oil Shock

The industrialized economies of the world were just regaining some growth traction in January 1979 when the shah of Iran, sick with leukemia, was drummed off his throne after months of riots and strikes, and was replaced by a gang of Shi'ite mullahs led by the Ayatollah Ruhollah Khomeini. A new and fearsome force was on the loose in the world: revolutionary Islamic fundamentalism. Throughout 1979, the Islamic revolutionary government of Iran choked down oil production to supply domestic needs only and stopped exports altogether.

Up until that moment, Iran had been the world's second leading oil exporter after Saudi Arabia, producing about 2.5 million barrels a day, representing about 5 percent of total world oil production, then roughly 50 million barrels a day (compared with 80 million barrels a day in 2004). Yet the loss of that 5 percent sent the global market price shooting up 150 percent—a classic case of system trauma provoked by a relatively minor tipping-point event. The Iranian revolution had sparked fear of Islamic fundamentalist trouble spreading to other oil-producing nations and prompted panic buying in anticipation of prices shooting up—which led precisely, in fact, to prices shooting up as demand suddenly surged.

The oil business had been changing rapidly since the United States had passed peak and lost pricing power to OPEC. Orderly procedures in the world markets had unraveled. Several major exporting countries had nationalized their oil operations—the Saudis, for instance, bought Aramco's assets lock, stock, and barrel by 1975—and distribution of oil and its products was no longer controlled vertically from wellhead to refinery to the gasoline pump as the old majors had done. The spot market replaced long-term contractual arrangements. Neither the new Alaska or North Sea fields could compensate for the market allocation disruptions. Panic provoked opportunism. Some OPEC members actually cut their production to tweak the spot markets even higher during the 1979 panic.

Prices at gas station pumps tripled. Gasoline lines returned to America. Inflation was rekindled as the price of making and moving everything in America increased again. The crisis went on a lot longer than the 1973 OPEC embargo. Even Great Britain, with its newfound oil wealth, rode the wave, boosting the price of North Sea oil to market levels. Though President Carter had warned the American public about a long-term energy crisis two years earlier, he was helpless to do much about the situation once it developed, except to write personal notes to the king of Saudi Arabia asking him to ramp up production to stabilize the price. Then, in the fall of 1979, a gang of Iranian "students" invaded the U.S. embassy in Tehran, Iran, calling it "a nest of spies," and took the remaining fifty-two American staff members hostage. The hostage crisis, which lasted more than a year, finished President Carter politically. Militant Islam's singling out the United States as the Great Satan among all the other industrialized nations launched a bitter culture clash that continues to this day.

Before the hostage crisis was over, Saddam Hussein of Iraq invaded neighboring Iran, an acting-out of age-old Arab-Persian religious schisms, border disputes, and political beefs. Among Saddam's first targets were the Iranian oil ports and refineries, which were damaged severely. The Iranians retaliated by shutting down all but one of Iraq's pipelines. The new war effectively removed a total of 8 percent of the world's oil from global markets and the price shot up again.

Glut

One consequence of the 1979 oil price boost was a boom in exploration and drilling with improved technology, especially deepwater sites. The assumption was that high oil prices were here to stay and everybody wanted to cash in. U.S. oil companies benefited as much as anyone from the high price level despite having passed peak domestically years earlier. (They were still pumping and still producing more than half of America's domestic needs.) The American companies plowed profits into exploring around the Gulf of Mexico and into new techniques for recovering oil from older fields. The Soviet Union, desperate to shore up its dysfunctional, decrepit, and bankrupt command economy, pumped as much oil as it could for the export markets. (In consequence, the Soviet oil endowment passed peak in 1986 and three years later the Soviet experiment collapsed.)

Meanwhile, the cumulative price shocks of two years running had sent the global economy into deep recession. Interest rates in the United States soared above 20 percent. Demand for oil fell as economies slumped.

By the early 1980s, the American gas-guzzler fleet had been replaced with much smaller cars, most of them foreign-made. The electric power industry was shifting massively to natural gas. Conservation practices were finally having an effect. The Persian Gulf countries had entered the robust phase of development and production, twenty years after their peak of discovery, just as Hubbert had predicted. Huge inventories from the panic buying spree of 1979–80 never sold down and much more new supply was now coming online. The Saudis had capacity to spare. The North Sea went into high gear. The combination of increased global oil supply and a moribund global economy took its toll. In late 1985, world oil prices collapsed. By early 1986 West Texas crude had plummeted from a high of $31.75 a barrel to $10. Some OPEC countries went as low as $6. The fifteen-year-glut was on.

The price collapse benefited many businesses, but hit U.S. oil companies hard. The majors quickly shut down exploration ventures and shed employees. Independents went out of business and suppliers went bankrupt as drilling stopped. Oil cities such as Tulsa and Houston went into economic tailspins. The new gleaming glass office towers of Houston were dubbed "see-throughs" because the occupancy rate was so low. American citizens, ever more dependent on their cars, were grateful for the reprieve on prices at the pump, and regarded the return of cheap gasoline as a restoration of normality. They were encouraged to think so from the top of American society. President Ronald Reagan was a cornucopian who believed that the oil supply was virtually limitless.

Energy soon fell off the U.S. political issues' radar screen. The American public figured that the crises of 1973 and 1979 had been a shuck-and-jive perpetrated by some mysterious "them"—politicians, oil companies, shifty Arabs, the Council of Foreign Relations, the Martians . . . they weren't quite sure. But as long as the price was back down at the pump, further concern faded. The cheap-oil industrial juggernaut could resume its march of progress. A world once again awash in cheap oil meant resurgent industrial activity and rebounding economic growth rates. Low oil and gasoline prices, in turn, stimulated demand in the United States, and with American oil companies languishing, increased imports fed the demand. Sensing a raw deal, the OPEC countries managed to put a quota system in place that would ramp up the average price to about $17 per barrel, and would form the basis for a global pricing system that held prices remarkably stable for another decade as virtually all oil nations reached production maturity. Saudi Arabia could always be counted on to act as the "swing" producer, using its huge excess capacity to modulate the global supply. The Saudis figured out that it was wiser to keep the Western industrial economies humming on moderately priced oil than lurching and staggering on higher-priced oil.

Iraq Attack and the Great Sleepwalk

The Persian Gulf War in 1991 represented only a minor deviation from what would end up being an extraordinarily stable decade. Saddam Hussein's Iraq had provoked a showdown with the United States by taking over Kuwait the preceding summer, supposedly because Kuwait had used snazzy new technology to drill horizontally across its border into Iraqi-owned oilfields. Once in possession of Kuwait, Iraq embargoed its oil. Prices more than doubled for several months, but the other OPEC states eventually compensated for lost market supply. Meanwhile, Saddam was demonstrably too crazy and grandiose to be allowed to keep Kuwait (and the 10 percent of the world's oil reserves it contained) so six months later, in the winter of 1991, the United States led an international coalition to evict him from Kuwait by force. In the process, a great deal of damage was wreaked on the Kuwaiti and Iraqi oil infrastructure by the retreating Iraqis. But as Operation Desert Storm was being conducted, the market price of oil actually fell back to a level below $20 a barrel, where it remained, with a few more minor deviations, through all of the 1990s and into the new century.

The 1990s represented an extraordinary final surge in the century-long oil fiesta. As the world approached the aggregate peak of production, a set of interesting conditions promoted stable prices in the face of greatly increased worldwide demand. Following the Desert Storm operation, the Saudis bent over backward to accommodate the United States. Whenever a supply bottleneck threatened to drive up prices, the Saudis, with their seemingly inexhaustible excess capacity, just opened the spigots another turn. The al-Saud family was thought to have been targeted by Saddam Hussein, who had dreamed of controlling the world's oil supply, and his defeat in Desert Storm had brought great relief to the kingdom. There would be nasty consequences to the al-Saud regime's "special relationship" with America, but they would not be manifest until early in the new century.

Meanwhile, South Korea, Malaysia, Thailand, Singapore, and especially China were becoming the world's manufacturing workshops as America "outsourced" heavy industry and focused its energies on hypertrophic suburban land development and the consumer infrastructure that went with it—malls, so-called power centers, and the vast highway strips with their fried-food shacks, tanning huts, and muffler shops. Americans were using more oil than ever before, and proportionately more of it was being burned in cars and trucks than for any productive activity. By the 1990s, American households were making a record eleven separate car trips a day running errands and chauffeuring children around. Automobiles were getting larger as the station wagon and van yielded to the supremacy of the sport-utility vehicle (SUV), an expeditionary car based on a light truck chassis and therefore exempt from legislated fuel efficiency standards.

In the 1990s, western Europe felt secure and the former Iron Curtain nations of eastern Europe emerged from the coma of communism. The North Sea fields were producing at full blast. Tourism had exploded in response to cheap air travel. The ethnic brawls of imploding Yugoslavia were an exception, and it is perhaps not insignificant that it was a geographic friction point between the Christian and the Islamic worlds. The former Soviet Union, or Russia, also contributed to the surge in production as its oil industry was reorganized on a quasi-capitalist basis. Attempting to get back on its feet economically, Russia's only means for obtaining hard currency was oil sales, as its decrepit factories produced nothing anybody would buy in an export market.

Apart from the Balkans, and the awful and innumerable civil wars of post-colonial Africa, and the low-grade insurgencies of Colombia and Peru, and the incomprehensible struggle in Sri Lanka, and the chaos of Chechnya, the nineties were a relatively quiet decade. The great proxy battles of the cold war were over. The United States and Russia did not even confront each other symbolically anymore.

With gas remaining cheap month after month, employment rising, and the computer revolution promising a "new economy," the American public entered a decade-long sleepwalk of complacency. Was it Bill

Clinton's immense good fortune to preside over two terms devoid of apparent crisis? If he had any reservations about the economy becoming hostage to the creation of suburban sprawl, he never voiced them. Like a lot of Sunbelters, he might have viewed sprawl as good to live in and good for business. Nor did he raise any alarms about the approaching global oil peak. He must have received intelligence briefings about it, even while the U.S. Geological Survey issued inflated estimates on total world reserves. The relative calm in world and domestic affairs from 1992 to 2001 led the public instead to a foolish preoccupation with trivialities such as the president's extramarital sex life and other celebrity misdeeds. As Bill Clinton yielded to George W. Bush in 2001, the only trouble on the scene was the disappointing slide in stock valuations and the extraordinary meltdown of Internet-based "dot-com" businesses that were supposed to form the infrastructure of the New Economy.

Then, one September morning that could not have been more beautiful in the eastern United States, nineteen Islamist maniacs hijacked four airliners and changed everything.

Crunch Time

As the United States waged war against "terror"—or what I call militant Islamic fundamentalism—in Afghanistan and Iraq following the 9/11 attacks, strange things coincidentally began to happen in the global oil markets. Princeton geologist Ken Deffeyes would contend that production data finally coming in during 2003 seemed to indicate that world oil had hit a production ceiling back in 2001, above which the world's producers could not penetrate.[11] Was this, in fact, *the* peak? The geologists in the Hubbert's Peak crowd knew that the peak would be detected only in the rearview mirror, several years beyond its actual occurrence, because the

11. Interview with Kenneth Deffeyes of Princeton with Julian Darley of Global Public Media (*www.globalpublicmedia.com*), April 4, 2003. "The *Oil and Gas Journal* publishes lists of country-by-country world oil production in the last week of each year and lo and behold in the years 2001 and 2002 . . . the numbers showed that the world oil production had not been as big as it was in the year 2000."

data took so much time to assemble. But these suspicious data suggested that something epochal had occurred. The price of oil was steadily going up, too, leaving behind the $20-a-barrel "ideal price" that the Saudis said would perfectly balance their need for profit against the West's need to maintain industrial growth, and therefore robust demand for oil. The markets seemed to know something.

By 2003, the price of a barrel of oil was ratcheting. It would go up three clicks and down two clicks, up four clicks, down two clicks, giving the appearance of just fluctuating a lot, but in fact heading up an inexorable trend line. In Venezuela, which accounted for 12 percent of America's oil imports, socialist President Hugo Chavez had wreaked havoc on the national oil industry by firing managers who opposed him politically. Production fell substantially and there was speculation that Venezuela, which had peaked in the 1970s, had such a battered, ancient, poorly maintained drilling infrastructure that production would never recover from the Chavez purge of expertise. In the spring of 2004, Shell Oil reluctantly disclosed in a report to stockholders that executives had misreported its reserves (oil available to be pumped) by 20 percent of the total, representing the equivalent of 3.9 billion barrels of petroleum, worth an estimated $136 billion. The disclosures caused a shareholders' uproar and led to a string of resignations. The scandal, however, was part of a much greater problem in the industry, which was that virtually all data on oil reserves from every company and every nation tended to be exaggerated on the high side.

By the summer of 2004 some additional strange things were happening. Demand for petroleum was rising implacably, especially because of continued industrial expansion in China, while capacity could not rise to meet it. China's oil imports doubled from 1999 to 2004 and surged nearly 40 percent in the first half of 2004 alone.[12] The Saudis and other members of OPEC stated repeatedly that they had a million barrels a day in surplus capacity, claiming in effect that they were still "swing" producers, able to dampen global markets at will by opening the valves a little bit wider. But despite their claims, production was not rising. Suspicion was rife that the mighty Saudi Arabia was just another entity that had misreported its re-

12. *Time* magazine, Asia Edition, October 18, 2004.

serves—and why not, since deep reserves conferred great political influence—and that perhaps the unthinkable had occurred: Saudi Arabia was actually hitting its peak fifteen years before the date even predicted by the Hubbert's Peak crowd (though it must be pointed out that they had based their calculations on the Saudis' own Aramco reserve data, as there was no other source of information about what lay under the ground in Arabia). There were whispers that Saudi Arabia's giant-of-giants Ghawar field on the Persian Gulf was catastrophically playing out, partly as a result of massive injections of seawater to keep pressure up. That same spring and summer of 2004, insurgents in Arabia ramped up attacks against foreign workers—who supplied most of the technical expertise in the oil fields. Housing complexes were bombed. An American trucker named Paul Johnson was decapitated in the most brutal conceivable manner by kidnappers and the gruesome event itself was played liberally over Islamic cable television, which sent more foreign technicians fleeing the country. Meanwhile, Iraqi insurgents were blowing up pipelines every other week, taking a million-plus barrels a day off the market for weeks at a time. On top of all these things, the Energy Institute in London reported that world production was declining by about over 1.25 million barrels a day as a result of depletion in twenty-seven oil-producing nations that were past their own individual peaks and experiencing sustained, inexorable production declines. In the fall of 2004, oil prices rose above fifty dollars a barrel.

Princeton University's Ken Deffeyes, a distinguished professor emeritus of geology, something of a jokester, but a fundamentally serious scientist, predicted on National Public Radio a few weeks prior to this writing that a final unequivocal global oil production peak would occur on Thanksgiving 2005, with "an uncertainty factor of only three or four weeks on either side."[13] This was a revision of statements he made in 2003 that the world had possibly already passed peak. The truth is that nobody will be able to

13. Deffeyes later explained his reasoning. From *www.MSNBC.com*, October 25, 2004: "How long will the world's oil last? Deffeyes argues that production capacity has grown more slowly than demand–based on production figures that are a lot more reliable than reserve data. 'Production is a pretty firm number,' he said. 'Oil gets counted twice: once when it gets produced and once when it goes into the refinery. So we pretty much know how much is produced, and my Thanksgiving Day prediction is entirely based on production.'"

state authoritatively when the world actually passes peak, except in hindsight, a few years after the fact. But we will know something happened because turbulence in the oil markets will be worse than what we are experiencing as I write, and military turmoil in various quarters will also reflect these dangerous circumstances. I think what Dr. Deffeyes may be trying to tell us is this: Let us give thanks for this extraordinary period of human history we lived through. Let us recognize that we are moving into a new phase of history. Let's be brave and wise about it, and prepare to move on.

Three

Geopolitics and the Global Oil Peak

In historical memory, the world has never faced such dangerous circumstances as it does early in the twenty-first century. The nations of the world face not only a life-and-death struggle over crucial energy resources, but an ideological struggle that makes the old capitalist/communist rivalry of the past century seem like a simple soccer match. Communism merely said, "We will bury you," and Comrade Khrushchev meant that in terms of economic and social progress. The avatars of inflamed Islam want to utterly destroy the infidel West, and its Great Satan seducer, the United States, and they mean down to the last beating heart.

This is a much darker time than 1938, the eve of World War II. The current world population of 6.5 billion has no hope whatsoever of sustaining itself at current levels, and the fundamental conditions of life on earth are about to force the issue. The only questions are: What form will the inevitable attrition take, and how, and in which places, and when? Some of these questions will be determined by the gathering calamity of climate change and its associated environmental implications, especially starvation, lack of fresh water, and the rise of epidemic disease (see Chapter 5). In the meantime, the world is faced by the dangerous posturings and maneuverings of nations around the control and possession of oil.

At the heart of this is the United States' sick dependency relationship with the Islamic world. Islamic nations possess most of the remaining oil in the world. We're addicted to that oil. Due to our inattention, narcissism, and almost unbelievably foolish complacency, we have allowed ourselves to become hostages to that addiction. We have enriched the ruling classes of the Middle East beyond the wildest fantasies ever dreamed by any emir, sheikh, pasha, or caliph in all the centuries of their dreaming. That wealth has transformed a poetic, decorous religion into a virulent agency of potential world conflagration.

From the point of view of a nation (America) struggling desperately to contend with it, much can be argued about the nature of Islam today and our peculiar relations with it. We have tried every imaginable tactic of political behavior modification to get along with the Muslim nations and all have so far failed. We have coddled them and bombed them, not necessarily in that order. We have spent untold billions helping them develop their mineral resources and finance public works. We have supplied them massively with armaments, some of which they are now using to shoot down American helicopters and stuff into car bombs, and we have sold them state-of-the-art military aircraft that we are prepared to destroy on the ground at a moment's notice before they can be used against us. At least one Islamic nation, Pakistan, incontrovertibly possesses nuclear bombs. The former director of its nuclear weapons development program, Abdul Qadeer Khan, is an Islamic fundamentalist who has been extremely liberal with the dangerous knowledge he possesses and in assisting others to get the equipment and materials necessary for nuclear bomb making. Another Islamic nation, Iran, ruled by mullahs who inveigh ceaselessly against America and its allies, is transparently building a nuclear industry that can be used for either electric power generation or bomb-making—take your pick. We have tried to be nice and we have tried to be harsh. Nothing works, and nothing is going to work. We are locked in conflict that can only result in chronic war until what is left of both America and Islam are forced to retreat back into geographic isolation on their opposite sides of the world.

For the moment, militant Islam is engaged in a jihad, or holy struggle, against the West and the United States in particular. Informed opinion is divided as to the motives of militant Islam and whether its behavior is psychotic or rational. George W. Bush is famous for claiming that "they hate us for who we are," especially for our freedom and democracy. Others say that this is nonsense—they hate us for what we do, specifically for our many military intrusions in Islamic lands, for our support of corrupt and dissolute regimes that tyrannize Islamic populations and squander their oil resources by selling too cheaply to the infidel West, for our support of Israel (which occupies a comprehensive realm of grievance all its own), and for our invasive degrading popular culture with its Satanic enticements to lewd and blasphemous behavior, especially among the young. To some, Osama bin Laden, the renegade Saudi millionaire revolutionary, is a psychopathic megalomaniac in the mold of Adolf Hitler, ingeniously bent on acting out his own twisted psychological life-script on the international stage and enlisting thousands of miserable suicidal dupes as his followers. Others say that bin Laden is a beloved righteous Islamic Robin Hood figure for a culture that hasn't had a hero to look up to for centuries, or until now enjoyed any victories over its many Euro-American oppressors. They add that bin Laden is following a strategy that is entirely rational in terms of Islamic theology, and tactically brilliant in its patient application of "asymmetrical" strikes against its enemies.

The rhetorical use of the term "war on terror" may be particularly unhelpful for Americans struggling to deal with these problems, as "terror" is not a nation or even a group or political entity, but rather a tactic in the service of an enemy that happens to be the widespread Islamic insurgency against American interests led by bin Laden and his al-Qaeda movement. Under Koranic law, jihad in the defense of Islam is the individual duty of every Muslim in the *ummah*, or worldwide Islamic community. As the United States has either troops, technical personnel, or intelligence agents in Afghanistan, Iraq, Saudi Arabia, Kuwait, Qatar, Turkey, Pakistan, Indonesia, Malaysia, the Philippines, and several former Soviet republics, it is hard to dispute the fact that we have a physical presence in Islamic territory—whatever our reasons—and that is enough to trigger jihad from the *ummah*, which has repeatedly issued explicit declarations of such. We are therefore at war with that community—not because of our choosing but because it has declared war on us. It is a circumstance that the U.S. government cannot possibly acknowledge honestly, for several reasons. One is that the government officials of many Islamic nations are effectively alienated from this *ummah*—by dint of their relations with us—and we are accustomed to understanding war only against governments and the armies they control, not against communities of faith, however broad they might be. In addition we are desperately beholden to these governments for our continued supply of lifeblood oil and we are bound to support them no matter how alienated they are from their own community—which only reinforces the jihadi rage and determination of the *ummah*. Another reason is our now pervasive political correctness, which forbids U.S. leaders from inveighing against another ethnic group or creed, or even acknowledging that deadly conflict may exist between the group and us. This leads to other forms of impotence, such as our chronic failure to respond to tactical insults such as the first World Trade Center bombing, the attack on the U.S.S. *Cole*, the Khobar Towers incident, or the bombing of embassies in Tanzania and Kenya, the attack on an international residential compound in Riyadh, the decapitation of American contract workers, and a long list of other hits, and our hesitation to engage wholeheartedly or to complete any military affray we find ourselves in, whether it is against Iraq in 1991 or Somalia, or giving the Taliban a month to escape from Afghanistan after 9/11, or the 2004 Shi'ite insurgency in Iraq.

While our intractable conflict with insurgent militant Islam has occupied center stage of the geopolitical scene for several years, we have many other concerns with other nations of the world—and they with us. Global relations have entered a period of turbulent flux, and events may turn in any number of strange directions before they resolve in what I believe will be the likely finale of the Long Emergency—with world powers retreating into their own regional corners, left to deal with fateful contraction of their societies due to the depletion of cheap fossil fuel. Whatever else happens, in the meantime we are still stuck with our sick dependency on Middle Eastern oil and the difficulties emanating globally out of it.

Peak and the Fate of Nations

Oil is the world's most critical resource. Without it, nothing works in industrial civilization as currently configured. Few people dispute the idea that the world will eventually run out of oil, and there is a broad recognition that it will happen sometime in this century—but there is next to zero understanding about what happens between now and then, on the way down to "empty." The American public also generally assumes that by the time the oil runs out humankind will have moved on to the next energy system—the current favorite candidate being one based on hydrogen—

and that it will arrive just in time, by special delivery, because the free market decrees it is so, and the free market never lets us down.

I don't believe it is going to work that way—and presently I will discuss the issue of alternative energy, the putative replacements for oil. The world will be in trouble long before we run out of oil, when we reach peak production. At absolute peak, there will still be plenty of oil left in the ground—in fact, half of all the oil that ever existed—but it will be the half that is deeper down, harder and costlier to extract, sitting under harsh and remote parts of the world, owned in some cases by people with a grudge against the United States, and this remaining oil will be contested by everyone. At peak and just beyond, there is massive potential for system failures of all kinds, social, economic, and political. Peak is quite literally a tipping point. Beyond peak, things unravel and the center does not hold. Beyond peak, all bets are off about civilization's future.

The public has been generally unconcerned or ignorant of the global oil production peak and what it portends. The previous oil shocks of 1973 and 1979 are viewed as transient difficulties that were overcome, and the illogical conclusion is drawn that all future problems with our oil supplies also will be overcome. The public has heard "experts" and "Cassandras" cry wolf about oil before—and no wolf appeared, life continued normally, so why take them seriously this time? Many people consider the peak oil story another fantasy brought to us by the same alarmists who said that the Y2K computer bug would bring on *the end of the world as we know it*. The attacks of September 11, 2001, were supposed to change everything, too, but we are still a nation of happy motorists tooling down the highway with our iced beverages and savory snack food products, with Rush Limbaugh cheering us along on the radio. Nobody is prepared for the sinkhole that awaits us down the road.

For many Americans, who have never known a way of life without cheap oil, there is a simple inability to imagine life without it. Some say that just because more oil hasn't been discovered doesn't mean that it isn't there. They are unimpressed by data showing that discovery peaked worldwide forty years ago and has been declining steadily since. In government, the discussion over our oil and natural gas situation has been plagued by misinformation, denial, and secrecy.

But it may be in the nature of crises that the conditions leading to them are ignored until it is too late to do anything about them. It may be hard to form a clear picture of a complex situation through a fog of facts and statistics. You decide. Here again are some of the salient facts of the global oil situation:

- The total planetary endowment of conventional nonrenewable liquid oil was roughly 2 trillion barrels before humans started using it. Since the mid-nineteenth century, the world has burned through roughly 1 trillion barrels of oil, half the total there ever was, representing the easiest-to-get, highest-quality liquids. The half that remains includes the hardest oil to get, lowest-quality liquids, semisolids, and solids.
- Worldwide discovery of oil peaked in 1964 and has followed a firm trendline downward ever since.
- The rate of oil use has accelerated tremendously since 1950. The explosive rate of world population growth has run parallel to our rates of oil use (in fact, oil has enabled the population explosion).
- The world is now using 27 billion barrels of oil a year. If every last drop of the remaining 1 trillion barrels could be extracted at current cost ratios and current rates of production—which is extremely unlikely—the entire endowment would last only another thirty-seven years.
- In reality, a substantial fraction of the remaining half of the world's total oil endowment will never be recovered.
- After peak, world demand will exceed world capacity to produce oil.
- After peak, depletion will proceed at 2 to 6 percent a year, while world population is apt to continue increasing (for a while).
- More than 60 percent of the remaining global oil endowment lies under the Middle East.
- The United States possesses 3 percent of the world's remaining oil reserves but uses 25 percent of world daily oil production.
- The United States passed peak in 1970 with the annual rate of production falling by half since then—from roughly 10 million barrels a day in 1970 to just above 5 million in 2003.

- The ratio of energy expended in getting the oil out of the ground to the energy produced by that oil in the U.S. oil industry has fallen from 28:1 in 1916 to 2:1 in 2004 and will continue falling.

The Bumpy Plateau

Looked at closely, the peak would resemble a kind of bumpy plateau because the price and demand data would all appear to wobble inconclusively for a while, perhaps for several years. High price, they say, "destroys" demand. As demand lessens, prices fall. Lower prices prompt demand to pick up again, and prices rise. The global peak period itself will be a period of both confusion and denial. Then, as the inexorable facts of the world peak assert themselves, and the global production line turns down while the demand line continues to rise, all the major systems that depend on oil—including manufacturing, trade, transportation, agriculture, and the financial markets that serve them—will begin to destabilize (including the oil industry itself). The peak will set into motion feedback loops of strange behavior as the boundaries among politics, economics, and collective paranoia dissolve, especially in relation to global markets and supply chains, which depend on a modicum of reliable expectations and transcultural trust. Once the world is headed firmly down the arc of depletion, fuel supplies will be interrupted by geopolitical contests and culture clashes.

Eventually, economic growth as conventionally understood in industrial societies will cease, or continue in only a few places at the expense of other places. On the bumpy plateau, global oil production rates may seem strangely at once robust and flat—robust because at peak the total barrels per day will remain in the highest historical range, but flat because they will never manage to go beyond a certain ceiling. Global production will never again increase. After oscillating at peak a few years, production rates will inexorably drop, and then the question becomes: how steep the drop?

During the singularity of the peak years, there will be no "swing" producer that can ramp up production in order to "flood the market" and keep global crude prices stable, as Saudi Arabia did for many years to the great benefit of the West. All producers will be pumping at the maximum rate. Rising demand among still-growing populations will bid up the price. The lack of a moderating market mechanism, such as surplus supply, to influence price will, by default, lead to allocation-by-politics. The politics of jihad (them) and blood-for-oil (us) will prove to be a very unfavorable basis for allocating scarce-but-indispensable commodities.

The economic stress among virtually all nations, the rich and poor, the advanced and "developing," will be considerable and is certain to lead to increasingly desperate competition for diminishing supplies of oil. Whatever a given nation's official take on the crisis may be, whatever level of denial or panic, all will be players in the ensuing contest for the remaining supplies of oil. The denial about global peak in the United States is already fierce, as investments in car-dependent, oil-addicted infrastructure are greater here than in any other nation and Americans consider their way of life a God-given entitlement. "The American way of life is not negotiable," vice-president Dick Cheney once famously remarked.

George W. Bush asserted in his 2003 State of the Union address that we would continue running that way of life on hydrogen. He did not indicate that the way of life itself might contain a few problems. Two months after that speech, the United States invaded Iraq in order to set up a police station in the fractious Middle East, which contains more than two-thirds of the world's remaining oil.

Flashpoint

The Middle East has been a geopolitical flashpoint for more than half a century. The first oil discoveries in Arabia were made in 1932 and it is not an accident that intense friction around the region has occurred in parallel with the development of its oil endowment. Without oil, the desolate Arabian peninsula would support a tiny fraction of the population now living there. It is otherwise devoid to an extreme of other resources, especially arable land and water. The modern state of Israel would not have been possible, either, without oil to fuel a modern European-type economy. It is a peculiar and tragic fact of history that the founding of the modern state of Israel occurred in tandem with the takeoff of the modern Arab oil states, and likewise, that the rise of Muslim fanaticism—as personified by al-Qaeda, the Taliban, the mullah-ocracy of Iran, Saddam Hussein, the suicide bomb death cult of the Palestinians, the militant sects of Pakistan and Indonesia—has focused its wrath on Israel as the embodiment of all Muslim hatreds for classic Western liberal civilization. Cairo, Baghdad, Kabul, Tehran, and other Muslim cities contained major Jewish communities in the late nineteenth century. Until the modern era, Islam had shown tolerance for these embedded populations, which comprised much of the commercial middle class. The evolution of Jewish settlements in Palestine into the nation of Israel over the twentieth century not only prompted a flight of these embedded populations to the attraction of a "promised" new homeland, but also instigated harassments and expulsions against the remaining Jews in those cities. As this process went forward into the mid-twentieth century, engendering hatred, envy, and resentment, the Muslim world took its cues for extreme anti-Semitism from the European totalitarian movements of the twentieth century—Nazism and Stalinism. All these tendencies have been further distorted and aggravated by the resource base of oil to which Western industrial society is addicted.

When Mark Twain visited the Holy Land in 1867, he found it weirdly vacant, and wrote in *Innocents Abroad:*

> [T]hese unpeopled deserts, these rusty mounds of barrenness, that never, never do shake the glare from their harsh outlines, and fade and faint into vague perspective; that melancholy ruin of Capernaum: this stupid village of Tiberias, slumbering under its six funereal palms. . . . We reached Tabor safely. . . . We never saw a human being on the whole route.

The industrialization of Europe across the nineteenth century provoked tremendous social change. On the one hand, the Jews of Middle Europe (Germany, Poland, Austria-Hungary, and others) gravitated to the booming cities, began to assimilate into mainstream culture, secularize, and enter cosmopolitan professions. On the other hand, new modes of transportation—railroads, steamships—made emigration easier, and large numbers of Jews chose to leave Europe altogether, many for America and a few for the sparsely populated Holy Land. The Zionist movement, led by Hungarian-born Theodor Herzl, institutionalized this epochal population shift into a nation-building venture. Herzl envisaged a wholesale transfer of millions of Jews out of Europe, both solving the age-old problems of anti-Semitism and fulfilling biblical prophecies of a return to Jerusalem.

The European Jewish pioneers who resettled this desolate land of biblical and Roman ruins in the early twentieth century arrived with up-to-date technologies in agriculture, irrigation, civil engineering, and other accoutrements of Western life such as electricity and telephones, setting the stage for a culture clash with Arab peoples whose manner of living remained stubbornly medieval. By 1914, approximately 100,000 Jews were living in the Holy Land, along with 600,000 Arabs. The Jews regarded the Arabs not unlike the way other European colonizers regarded their colonial natives, with condescension. Arabs there (in a region not yet called Palestine) nonetheless benefited from the area's economic development. Some yielded to modernity, became urban and educated, and established themselves in commerce and the professions, which would lead inevitably to politics. Their population grew.

Meanwhile, World War I broke out in Europe. The mass slaughter and futility of the trenches went on for years and national treasuries groaned under the strain. Both the British and the Germans hoped to gain aid from their Jewish financial networks to continue paying for the war. One result was the Balfour Declaration of 1917, in which Britain essentially took up the cause of Zionism. Another result would be the postwar German resentment at being "stabbed in the back," as Hitler would put it, by Germany's Jews.

At the war's end the geopolitical remnants of Germany's defeated ally, the Ottoman Empire, had to be disposed of, including its possession of the Holy Land, now de facto a pan-European Jewish outpost under British control. The British engineered a quasi-colonial protectorate under a League of Nations mandate in what would now be called Palestine. At the same time they established the kingdom of Transjordan by plucking a Hashemite chief named Abdullah out of Arabia and installing him on a throne in a previously unincorporated zone east of the Jordan River, with the expectation that it would become the de facto homeland of the majority of Arabs in the region—that is, the people who today call themselves

Palestinians. The British underestimated the Palestinians' attachment to Zion proper.

Once Palestine was more or less officially established as a Jewish homeland, yet more purposeful settlement occurred in waves, first by those who saw clearly what Hitler intended before World War II, and then, of course, of those who survived the Holocaust and arrived later. Just before and then after the war, the oil resources of Arabia, under the Saudi royal family, went into development, first by the British and then by American companies. Oil would soon bring unheard-of riches to a people who had lately been a sparse tribe of camel-riding nomads in an ocean of sand and rock. The establishment of Israel as a sovereign nation in 1948—an ethnic anomaly along a Muslim strip nine thousand miles long reaching from Morocco to Java—provided the Islamic world with a focus for the anxieties that oil unleashed, amplified by the difficult adjustment to sudden enormous wealth and the cultural incursions of modernity that came with it. And this was occurring just as the European nations were systematically dismantling their obsolete colonial empires and withdrawing back across the Mediterranean, leaving expanded Muslim populations to govern themselves.

Another factor in setting the tone for chronic Arab-Israeli strife was the refusal of Muslim states around Israel to absorb Palestinian Arabs displaced after the struggles that attended the founding of Israel in the late 1940s. The Palestinians would become pawns and proxies both of the cold war Soviets and of their Islamic cousins in a series of battles lasting half a century that has never been resolved nor ever quite added up to a major regional war lasting more than a couple of weeks. It became the role of the Palestinians to act out all the Muslim world's real and imagined injustices. Had they been absorbed into other Arab countries, those nations would not have so easily diverted themselves from their own internal contradictions.

The major fault lines in today's Arab-Israeli struggle were first established by the charismatic postwar Egyptian leader Gamal Abdel Nasser, a colonel who seized power from the feckless King Farouk in 1952. Nasser hoped to use military aid from the Soviets to unite all the Islamic nations of the Middle East into a single entity, a United Arab Republic, and in the process to destroy Israel. However, only Syria enlisted, and only until

1961. Nasser took it upon himself to act out many of the power fantasies that the Saudis, with their addiction to oil revenues and dependence on Western expertise and markets, could not dare to. He was among the first to propose that the Islamic world use oil as a weapon against the West, though Egypt itself possessed negligible amounts of the resource.

But Nasser was also a secularist who found religious fanaticism politically inconvenient. He outlawed the budding Islamist movement called the Muslim Brotherhood, which took umbrage at, among other things, Nasser's romance with Soviet-sponsored socialism because it violated the Koran's sacred notions about private property. Nasser arrested the brotherhood's intellectual leading light, Sayyid Qutb, and eventually hanged him, but not before Qutb had produced reams of antimodern manifestos, including a thirty-volume exegesis of the Koran—the thrust of which was to propose the imposition of Koranic law, shari'a, on all the nations of the Islamic world—in other words, the establishment of theocracy rather than secular military junta à la Nasser.

Many of the Muslim Brotherhood's leading intellectuals—including Qutb's brother Muhammad—were sent fleeing from Cairo and found refuge in Saudi Arabia, where they were welcomed by the arch-conservative Saudi princes, who were desperately short of scholars, swimming in money to pay them, and eager to fortify their branch of Wahhabi religious practice with intellectual respectability. In an extremely bold move in 1956, Nasser seized the Suez Canal from the Anglo-French consortium that ran it. The Suez Canal, connecting the Red Sea to the Mediterranean, was a crucial link for tankers delivering the bulk of Europe's oil from fields in Arabia, Iran, and as far away as the old Royal Dutch concessions of Indonesia.

Nasser's seizure of the canal was ostensibly retaliation for the United States and Britain's decision not to finance the Aswan Dam project on the Nile. That decision, perhaps foolish in hindsight, was based on Nasser's flirtation with socialism and his cozying up to the Soviet Union, which had supplied Egypt with massive amounts of advanced weaponry, including scores of MIG fighter-bombers, in an attempt to promote conflict with Israel. This was, of course, part of the greater conflict between the Soviets and the West over the Middle East. The Soviet goal was to do everything possible to make the nations of that region inhospitable to free Western influence and hegemony.

If the Aswan Dam decision was a blunder on the part of the West, then Nasser's blunder was to blockade the Israeli port, Eilat, on the Red Sea. The Egyptians had been harassing Israel for years by sponsoring Palestinian terror operations out of Gaza, Jordan, and Syria. The French and British now convinced the Israelis to use the blockade of Eilat as a justification to invade the Sinai and put an end to these shenanigans. The plan had been worked out secretly among the three in advance. Britain and France would enter the fray, ostensibly to separate the Egyptians and Israelis, then establish a ten-mile buffer zone on each side of the Suez Canal, and regain control over it, thus accomplishing three strategic goals: (1) to assure the continued unimpeded transport of oil supplies to Europe; (2) to thwart Soviet attempts to gain de facto control over Suez; and (3) to put a damper on Nasser-style militarism.

Israel did indeed invade the Sinai on October 29, 1956, and effectively secured it after one hundred hours of fighting. But the larger plan failed when the United States declined to support Britain and France. The Soviets, meanwhile, threatened to intervene using "every kind of modern destructive weapon" in support of Nasser. President Dwight D. Eisenhower feared a sudden rush into World War III. He demanded that the British and French back off and threatened to cut off economic aid to Israel if they did not quit the Sinai. The United States then induced the United Nations to call for an immediate cease-fire. Eisenhower and his secretary of state, John Foster Dulles, also saw benefits in currying favor with the postcolonial "third world" nations, which were then coalescing into an influential voting bloc in the UN (disproportionate, perhaps, to their actual power but giving them the ability to obstruct many UN actions)—not to mention the flat-out competition with the Soviets for their hearts and minds.

In the years following the 1956 Suez crisis, Arabs and Israelis maintained an uneasy truce. Tensions escalated again in the mid-1960s as Arab leaders jockeyed against each other for dominance in the Middle East, using anti-Israeli posturing as a way of distracting their populations from the internal social and economic problems they faced. Egypt's Nasser struggled to maintain his position as the preeminent figure, but his United Arab Republic idea fell apart. The Syrians increased their border harassments against Israel in 1966 as the Baathist Hafiz al-Assad encountered economic problems in his oil-poor nation.

The Palestinian Liberation Organization (PLO) was midwifed in Cairo in 1964 under Nasser's direction, along vaguely Marxist revolutionary lines, Marxist "liberation" politics being the fashion of the time. The PLO's charter explicitly called for the use of violence for the purpose of expelling the Jews from Palestine. In 1965, the Fatah arm of the PLO began terrorist attacks against Israeli civilians. The PLO eventually managed to instigate the gravest rift in inter-Arab relations in the early 1970s when their armed gangs, using Jordan as a base, threatened the civil order of that country. Eventually, King Hussein massacred several thousand Palestinian irregulars, which led to a rash of assassinations, hijackings, and terrorist exploits under the rubric of the Black September movement. All the while, the Soviets and the Americans competed in supplying arms to their clients in the region, creating the dangerous conditions that led ultimately to the Six-Day War in 1967 and the Yom Kippur War in 1973, described in the previous chapter.

Against this background, we move to the present situation. Israel has become the region's sole nuclear power. It operates a fleet of submarines armed with nuclear missiles that roams all over the globe, impervious to threats. Israel is said to possess at least two warheads programmed to strike each of the capital cities of its enemies. Egypt has long ceased to be a military threat to Israel. The Soviet Union no longer exists as anyone's patron. By default, Egypt has become the financial ward of the United States. Egypt and Israel officially still observe the 1978 Camp David peace accords, but the status quo depends on the continued rule of Hosni Mubarak and not much else, and he has begun to show signs of failing health in his mid-seventies.

Israel has been indispensable to the Arab world as the default distraction from local or domestic issues, the stand-in for all of Islam's quarrels with the outside world. Without Israel to focus on, all the anxieties and political frustrations of many Muslim states would turn inward, in the form of revolution, insurrection, and civil strife. Paradoxically, a strong Israel is more valuable to the current generation of Arab politicians than a vanquished Israel would be.

Problematic as it may have been for the United States as an ally and client, Israel has been indispensable as the regional policeman of last resort. The Israeli intelligence service in particular has been a tremendous benefit to the United States as a window on the dark interstices of Arab politics. But in the new age of asymmetrical warfare, the constant deadly harassments of terror acts, suicide bombings, and armed incursions against civilians threaten the viability of "normal" life in Israel. A negotiated resolution of the Arab–Israeli conflict will not happen as long as the neighboring states enjoy the artificial prosperity of oil wealth. For that matter, the modern state of Israel, being a creature of the industrial era and as dependent on oil as any industrial society, may not survive the fossil fuel crises of the coming decades. The exploding Palestinian population itself might be the ultimate weapon that would overwhelm the experiment of a modern Jewish state, but as the oil runs out, the region will probably not support continued population growth by any group. Life will become much more desperate when the struggle for resources intensifies, and it is not hard to imagine circumstances that would turn Palestine back into the kind of sparsely populated wasteland that Mark Twain encountered only 150 years ago.

For half a century, the Arab-Israeli conflict has been a mask over the much graver contradictions of the West's ever-growing dependence on the oil resources of a handful of Arab and other Muslim nations in the Middle East (e.g., the Persians of Iran). Whatever else was going on in that region, the United States and Europe have enjoyed, with few interruptions, a remarkably steady supply of the single most indispensable resource to industrial civilization, and at favorable prices—especially once the United States passed its own oil production peak in 1970 and lost its role as the global swing producer that could push prices down by simply pumping more. Plain avarice and a lack of other options had this handful of Middle East oil nations in a thrall of codependency with their addicted patrons. In a very few years all the other oil-producing nations of the world will be past their individual peaks. This will leave Saudi Arabia, Iraq, and Kuwait in a special and uncomfortable position, with an energy-starved world of armed and dangerous nations glaring hungrily at them (and menacingly at each other). Then these oil-rich few will face the prospect of their own inevitable depletion. The tensions arising from these linked prospects have, I believe, contributed hugely to the pronounced mood of political animus currently gripping the Islamic world.

That world has entered a period of the most extreme turbulence, from Algeria to Pakistan, extending through the old frontiers of the former

Soviet Union and down into Indonesia, the world's most populous Muslim nation. And this turbulence coincides with the climactic phase of the Arab oil bonanza. It is a multidimensional turbulence—religious, ethnic, ideological, economic—and it is taking place on an underlayer of ecological desperation, as populations in many Muslim nations grossly overshoot the carrying capacities of the places they inhabit.

Even nonreligious observers must regard with awe the potential that the Middle East now holds for setting off a civilization-ending war, a virtual Armageddon. The approaching end of the oil age seems to have propelled the Islamic world into an accelerating mood of dark, bottomless fury as it sees its wealth squandered, looks out on its seething populations, and views a harsh and hopeless post-oil future.

The Main Event

By the time you read this, Arabia may no longer be the personal property of the al-Saud family. This was certainly the opinion of Robert Baer, twenty-three-year veteran of the CIA's Middle East bureau and author of *Sleeping with the Devil,* a comprehensive glimpse into the recondite doings of the al-Sauds.[1] The kingdom, which did not exist as a formal nation until the 1930s (not much longer than the Israel it detests and envies), and may not exist much longer as a kingdom, carries on as one of the most brutal and repressive police regimes of the post-cold war era. Corrupted to the core by a half-century of unearned mega-riches, trapped in the practices of the most severe branch of Islam (Wahhabism), cravenly cowed into financing the very terrorist network intent on the ruling clan's destruction, the al-Sauds run Arabia as the Bourbons ran France in 1789, literally as if there were no tomorrow.

Arabia, Saudi or otherwise, holds 25 percent of the world's remaining oil reserves. (The Middle East as a whole holds at least 60 percent of the world's remaining known reserves.) Depending on which sets of statistics you chose, Arabia will pass its own oil production peak sometime between 2001 and the year 2020, meaning that in theory it could continue

1. Robert Baer, *Sleeping with the Devil,* New York: Crown, 2003.

as the world's producer of last resort for quite a while. But that is assuming a lot, starting with the actual oil reserves themselves. Reports by the Arabian national oil company, Aramco, are themselves highly suspect, because the actual statistics are considered state secrets. Overstating reserves has been endemic for years, as OPEC production quotas have customarily been based on how much oil each country possesses, and the higher the estimate, the more each OPEC member has been allowed to pump out of the ground and sell.

Arabia's surplus capacity made it the epicenter of the oil world for a generation. There are reasons to suspect that Arabia may have already entered its peak phase. By 2003, with Iraq's oil industry in a shambles, Venezuela crippled, Nigeria intermittently paralyzed by civil disorder, and the North Sea depleting at around 5 percent annually, Arabia was rumored to be pumping at full capacity. It had reached its capacity "ceiling." It no longer had the option of opening the valves a little wider to stabilize the price of crude on world markets. It was no longer the crucial swing producer. As Kenneth Deffeyes remarked at the time at an international conference on the global oil peak (I paraphrase), the good news was that OPEC could no longer dictate world oil market prices; the bad news was that no one was in control any longer. By the summer of 2004 there were further signs of deterioration. With the prices soaring toward $50 a barrel on the New York markets, the Saudis promised repeatedly to increase production and consistently failed to do so. In fact, during August 2004, daily production actually fell by about half a million barrels.

This telling behavior was accompanied by physical evidence of the approaching depletion of the Arabian oil fields. Only six giant oil fields account for all of Arabia's production. The largest is Ghawar, the biggest oil field ever discovered, a 300-mile-long sliver-shaped piece of real estate straddling the kingdom's Persian Gulf shoreline. It accounts for 60 percent of all the oil ever produced in the kingdom, and currently accounts for 5.5 percent of the world's daily production. In 2004 it showed ominous signs of falling into depletion.[2] For years, Aramco, the Saudi national company, has injected seawater into Ghawar as a method for forcing the

2. Institute for the Analysis of Global Security, "New Study Raises Doubts about Saudi Oil Reserves," March 31, 2004, *www.iags.org*.

oil out under pressure. It is currently injecting a staggering seven million barrels of seawater a day. As this occurs, over time, an increasing amount of the outflow is composed of water along with the oil. By the summer of 2004, this so-called "water cut" was estimated to be up to 55 percent of the outflow, meaning that more than half the liquids flowing out of the Ghawar fields were water. It is estimated that Ghawar's production is already dropping by 8 percent a year. To compensate for that, seawater injection will have to increase, along with advanced extraction techniques such as horizontal drilling and "bottle brush" drilling, to make up for the decline. These methods do temporarily boost production, but they lead to accelerated rates of depletion in the future. They also have the effect of damaging the underground geological structure of the oil field. It is likely that Ghawar's reservoir may have already suffered damage that will lead to a shorter lifespan than had been previously expected. Aramco claims that Arabia's ultimate recoverable reserves amount to 250 billion barrels. Other authorities, including American oil companies that originally explored the peninsula and ran the drilling operations there until the al-Sauds kicked them out in favor of Aramco, estimated ultimate recoverable reserves at 130 billion barrels. It is estimated that total oil produced in the nation since drilling began stands at 100 billion barrels. If you only split the difference between the two figures on ultimate recoverable reserves, Saudi Arabia would be past peak.

A lot of the management of Aramco's daily business has been performed by the roughly 100,000 expatriate guest workers. About 65 percent of these are British and American, the remainder Japanese and other nationalities. Deadly attacks on foreign worker living compounds in the spring and summer of 2004, coupled with the horrifying video-recorded beheadings of foreign hostages such as American truck driver Paul Johnson, are an obvious tactic by radical Islamic insurgent forces (al-Qaeda or otherwise) to drive oil production expertise out of the kingdom. As George Friedman of Stratfor.com has pointed out, it is easier for insurgents to achieve their goals by targeting guest workers than by blowing up oil wells, pipelines, refineries, and terminals. "Attacking oil assets also attacks the Saudi gravy train—oil is 90 percent of Saudi Arabia's export revenues. . . . Attacks against expatriates are akin to slowly turning the screws on Washington and Riyadh, while attacking the infrastructure directly is a sledgehammer blow that al-Qaeda can only use once."[3] Strategically, opponents of al-Saud family rule want the oil to be available to whatever regime succeeds the royal family. Yet it is probable that insurgents will retain the option of striking it selectively some time in the future. In the meantime, driving expatriate guest workers from the kingdom under threat of beheading, bombing, or shooting would have the effect of degrading the production capacity, starving the al-Saud princes of needed revenue, and producing economic disarray in the West, America in particular.

The reason I have made a distinction between Arabia and Saudi Arabia is that sooner or later the Arabian peninsula will cease to be the personal possession of the al-Saud clan. When that happens, we will be dealing with a very different Arabia. Arabia is the center of gravity of the Islamic world and the world oil markets. In Baer's phrase, Arabia is "the fulcrum on which the global economy teeters." American policy for the past quarter-century has been based on the delusion that Saudi Arabia is stable and that America can enjoy regular supplies of its oil at a fair price indefinitely.

Saudi governance is already imploding into its own internal vacuum of extreme cupidity, self-deceit, sloth, apathy, and inertia. The modernity that the 30,000-odd al-Saud family members found themselves in, especially at the highest royal levels, was a purposeless modernity of dissipation. The thousands of princes on the oil dole have received welfare benefits—anywhere from $19,000 to $270,000 a month—unmatched by any royal family at any time or place in history, as well as a steady income of "commissions," bribes, and other kickbacks in the construction industry and arms deals. Their positions have also permitted extremely liberal expropriations of other families' private property, a kind of blank-check eminent domain system for the benefit of the al-Saud clan members only. They have squandered much of these immense fortunes on pleasure-seeking, toy-buying, jet-setting, and self-aggrandizing land development. (Prince Abdul Aziz, a favorite of the disabled King Fahd, built his own $4.6 billion theme park featuring scale models of Mecca and other Islamic holy places, with actors carrying on within.) At least a trillion dollars of Arabian money has been parked in American securities markets for the

3. *Stratfor.com*, June 9, 2004.

past decade; the rapid withdrawal of it could bring American finance to its knees (and probably global finance with it).

Because the Arabian birth rate is among the highest in the world, the annual oil welfare allotment of ordinary Saudi subjects—that is, non-royal family members—fell from a high of $28,600 in 1981 to less than $7,000 in 2003. Seventy percent of all jobs in the kingdom and 90 percent of private sector jobs are filled by foreigners.[4] The depressing bill of particulars about the daily details of life in the feudal kingdom is a prescription for anomie of the most extreme kind. Few Arabian national adults work at real jobs. The sexes are completely segregated. The temperature is over 100 degrees half of the year. The nation leads the world in beheadings. Higher education is free, but half the Ph.D.s awarded are in Islamic studies. Liquor and movies are forbidden (the ruling elite can do whatever they like within their private compounds, or travel free on the national airline to places where such recreations are allowed). Unemployed young men have little to do but contemplate the futility of their prospects—and the overthrow of the regime that maintains the status quo.

The very religious doctrine that the al-Sauds have been financially underwriting for decades, and the enormous infrastructure supporting it—from the thousands of madrassas, or Islamic academies around the region, to the al-Qaeda training camps, to mosques in the United States and Europe—is dedicated to extirpating exactly the kind of decadence embodied in the behavior of the royal family. The kingdom has huge numbers of young men without jobs or prospects of jobs and the religious schools have been the chief means of occupying them. Clerics inside Arabia openly call for jihad against all infidels, by clear implication including the ruling family. The al-Sauds have been pouring money into this enterprise of their own self-destruction in the hope that they might divert fundamentalists' attention away from their own wretched excesses to the sins of the West, and of hegemonic America in particular, and so far it has worked.

The chief concern from the point of view of America and the rest of the industrialized world is that a revolution against the al-Saud clan would

4. Robert Baer, "The Fall of the House of Saud," *Atlantic Monthly*, May 2003, p. 58.

likely be carried out by these fundamentalists—as opposed, say, to democratic reformers, socialists, competing royals, or any other conceivable rivals to power—and a further implication is that such a revolutionary government in Arabia could be headed by the world's leading Arabian Islamic radical, Osama bin Laden, or someone like him. This would almost certainly lead to a cessation of direct Arabian oil imports for the United States. Since oil is fungible, meaning a batch of light crude from one region can be traded for a batch of light crude in another region, the United States might be able to work around such a boycott in world oil markets. But a revolutionized Arabia could simply reduce total production. In fact, it would be in its interest to do so, as oil is the only exportable resource it possesses and the nation would be far better off husbanding whatever it has left. What's more, fundamentalists in power, unlike al-Saud princes, would not require billions of dollars in spending money for the maintenance of yachts and trips to Monte Carlo, so they could more easily afford a necking-down in production. Disburdened from supporting several thousand decadent princelings, a revolutionary government could reduce oil production and still maintain subsidies for ordinary Arabian citizens. This assumes, of course, that the infrastructure of oil production would not be damaged in the first place during a revolutionary struggle. Were al-Qaeda to direct such a revolution, its ultimate strategic goal would be the establishment of a pan-Islamic *ummah* with Arabia at its center, containing Islam's two holiest sites, Mecca and Medina.

Of course, the global industrial economy would not easily tolerate a significant reduction in Arabian oil exports, nor an Arabia as a permanent jihadi base, and any number of possible scenarios might be spun out of that circumstance. One is that the United States might attempt to intervene in support of a threatened Saudi regime, including an attempt to physically take over and secure the Aramco oil assets on the ground, a gambit that could fail spectacularly. Destruction of the Arabian oil infrastructure has been a nightmare for U.S. government strategic planners, second only perhaps to the spread of nuclear weapons. From the insurgents' point of view, the system can be attacked at a few key points using weapons and explosives easily obtained in the international arms bazaar, and hitting some select targets would afford them tremendous leverage against their adversaries at low cost.

The Alqaiq refinery, the world's largest, processes more than half of the nation's crude. From there it is piped thirty miles or so to a couple of terminals on the Persian Gulf coast, Ras Tanura and Ju'aymah, where it is loaded on tankers at the offshore facility called the Sea Island. The terminals are heavily defended against external attacks, but Saudi intelligence, the army, the police, and even Aramco have been infiltrated by bin Laden sympathizers and it would be next to impossible to stop a determined insider bent on sabotage within the facilities themselves. Pipelines running from the giant oil fields of the Persian Gulf west to Yanbu'al Bahr on the Red Sea are indefensible. Most of that oil goes through the Suez Canal to Europe. The pipeline could be taken out with a camel and a few pounds of Semtex explosive at any point along a five-hundred-mile transit. Action against wellheads, of which there are thousands, could cause irreversible damage to the oil fields themselves. If the Saudi oil infrastructure were crippled, the global economy would stagger, with the U.S. economy leading the way off an economic cliff. Normality, as it has been understood in the United States for a long time, would end very quickly.

Many Arabians regard oil as a curse. They have lived with this bonanza a little more than half a century. It has turned their lives upside down and inside out and has devastated their traditional culture. An Arabian proverb of our time goes something like this: "My father rode a camel, I drive a Rolls-Royce, my son flies a jet airplane, and his son will ride a camel." The fatalism is revealing. How will Arabians live after the oil bonanza ends? Many in the generations under forty years of age have never known life without air conditioning, automobiles, and shopping malls. Even under the most favorable circumstances, the Arabian oil endowment would not last more than another fifty years. There will come a time—before the end of the twenty-first century—when circumstances will compel a return to traditional lifeways. The prospect is sobering. The region will not support anywhere near the number of current inhabitants, who owe their existence in one way or another to the subsidy of their oil endowment. There must be many Arabians who follow this thread to its logical conclusions and feel the terrible weight of destiny bearing down on them. It is their palpable and inescapable fate—and why wouldn't the apprehension of this destiny not propel a demoralized people into a desperate religious hysteria?

Of course, it is for the Arabians to sort out what they will do in this fast-approaching future. But the United States will face a quandary in the even shorter term, if confronted one way or another with the precipitous cutoff of its Arabian oil imports, by either sabotage or politics. America has demonstrated its willingness to invade sovereign nations in the Middle East—and ultimately justifies it on the basis of the Carter Doctrine, which states that oil supplies represent vital U.S. interests to be defended by military force if necessary. In regard to the problems presented by Arabia, there are two basic questions for us: What might the United States do? And what *can* it do?

Another Hegemon

The United States might attempt to occupy the Arabian Peninsula—perhaps including Kuwait, Qatar, and the United Arab Emirates. It's possible that we could accomplish the first stage of a military takeover, but doubtful whether we could effectively maintain that presence; in that case, what would be the point? The clear lessons of Iraq are that countries may be easy to roll over but not so easy to pacify afterward. There is only so much territory that even a superpower can hope to control at any given time. There are limits to the number of U.S. troops available for deployment, and to the appetite of the American public for war and sacrifice. There are only so many financial resources available to support military adventures. But a desperate superpower might feel it has no choice except to attempt to control the largest remaining oil reserves on the planet at any cost. In this event, Arabians might have more incentive to sabotage their oil infrastructure. As outlined above, the prospects are poor that we could defend it, in which case there might be no benefit to occupying Arabia, except perhaps to prevent another power from forming a hegemonic relationship with the region. And who might that be?

China will need petroleum at least as desperately as the United States in the years ahead. China has little petroleum of its own, and it has been explored relentlessly, acre by acre, as a purely government enterprise, unhampered by normal cost considerations. The communist party sent drilling crews absolutely everywhere, with disappointing results, and there is nowhere left to explore except offshore in the South China Sea, a region contested by several other nations. China has even less oil than the United States has left, with four times the population. China has ramped up an industrial economy that is now the world's second greatest consumer of oil, surpassing Japan in 2004. In fact, it could be said that China has launched the last industrial economy of the oil age, and gotten it under way too late in the game. China's oil imports doubled over the past five years and surged nearly 40 percent in the first half of 2004 alone. At the current rate of growth in demand, China alone, of all the world's nations, will consume 100 percent of currently available world exports in ten years—assuming no growth in demand elsewhere in the world and assuming no falloff in global production.

China is a nuclear power with a nuclear "umbrella" that it can spread out to shelter client nations. China is geographically closer to the Middle East than America is, and might shield Arabia at least as effectively as America has for half a century. China could enter into a protective relationship with any number of nations from Central Asia to the Middle East, including an Arabia run by a militant Islamic theocracy. In short, with China becoming a presence by necessity in the region, we would be back in a cold war again, or something worse, contesting with a rival world hegemon, this time over energy resources, not ideology. What's more, China could enter into a protective relationship with a nation like Arabia without any of the West's religious encumbrances as historic infidels and "crusaders." It isn't a huge stretch to imagine this new cold war heating up when people start shivering and going hungry in Ohio.

The China-ascendant scenario is predicated on the probable failure of the United States to control the Middle East militarily for an extended period of time. We find ourselves now engaged in what amounts to the opening ceremonies of a third world war against one billion angry Muslims, with forces enlisted from Islamic nations outside the main theater of operations, and with China poised on the sidelines waiting to pick up the pieces. Such a widescale conflict, conducted asymmetrically, raises the specter of terrorism on the grand scale, possibly with nuclear weapons or "dirty" radioactive bombs, and retaliations in kind. The ultimate result would be a retreat by an exhausted and bankrupt United States to the Western Hemisphere, to severe economic discontinuity and internal political strife.

In the chaotic period around the peak oil event, China will not be without extraordinary problems of its own, starting with enormous population pressures, and moving on to massive environmental degradation and the incubation and spread of epidemic diseases, including deadly influenzas associated with factory farming as well as accelerated AIDS infection (see Chapter Five). The severe acute respiratory syndrome (SARS) outbreak was a preview of coming attractions. On top of these vicissitudes will be added the severe economic hardship entailed when an economically strapped America (and the rest of the West) can no longer sop up the many products of China's tremendous industrial capacity. This would produce widespread unemployment in China, possibly leading to political turmoil of a kind not seen since the Cultural Revolution of the 1960s.

Those of us who lived at the time of the Cultural Revolution must remain impressed by China's potential for lapsing into political psychosis. China's material progress in the past four decades has also been impressive. This progress coincided with the global oil fiesta now culminating in the production peak. When that enabling mechanism is withdrawn by historical circumstance, China may not hold together. If it implodes in political chaos, there is no telling what will happen with its neighbor (and historical enemy) Japan. Japan has even less oil and natural gas than China, but it has a much smaller population and a far more disciplined and rigorous social infrastructure. It might be subject to nuclear blackmail. It might be drawn into the violent vortex of China's meltdown by other means. It might find ways to persevere and even reassert its dominance over this part of the world when other nations fall apart. Japan could also retreat into the hermetic isolation of its own islands, as it has done at other times in history, letting China do to itself what it will.

Iraq and Iran

What was the war in Iraq about? It was strategically about setting up a police station in the middle of a very large bad neighborhood. It was also about dividing the Islamic world physically in half to create a buffer between the aggressive gangs on the east side of the police station (Iraq, Iran, Pakistan) and the politically touchy gangs on the other side (Arabia,

Egypt, Syria, Yemen, Libya, and, of course, the anomaly, Israel). On a closer scale, the Iraq war was an attempt to establish a forward base adjacent to Iran and Arabia, to moderate and influence the behavior of both of them, to discourage adventures by Iran and to be ready in case of trouble in Arabia. One of the first things the United States did after invading Iraq in 2003 was to station two armored divisions on the Iraq-Saudi Arabian border.

Then there was the issue of Iraq itself. Iraq possesses the world's second-largest oil reserves, after Arabia. Replacing the Saddam regime with a less hostile and erratic government was obviously a high order of business for the United States where future oil supplies were concerned, but there were other compelling strategic considerations. The ostensible reason for the war—that Saddam Hussein possessed nuclear, biological, or chemical weapons of mass destruction (WMDs)—was ultimately proven erroneous, but to label it a mendacious ploy is unmerited. The precipitating justification for the war was Saddam Hussein's refusal to let UN inspectors visit all the sites they deemed necessary to look through. Among these were a dozen extensive underground bunkers engineered by German contractors during the 1980s. The bunkers had been labeled "presidential palaces" by Saddam and placed off limits to UN inspectors. The bunkers represented hundreds of thousands of square feet of secure, underground storage space, immune to satellite photography or other advanced spying technology, where any amount of WMDs might have been stashed—fissionable material, nuclear devices, stocks of anthrax, smallpox, chemicals. The available intelligence about WMDs might have been flawed, dubious, or unreliable—but most of all it was inconclusive. The United States needed to know conclusively if anything dangerous existed or had left traces in these places and possibly had been shifted elsewhere, perhaps out of the country. Since the UN team was prevented from completing the search, the United States had to do it in person. The fact that nothing was found by American forces after the 2003 invasion does not prove that we didn't have to look.

There were other good reasons for getting rid of Saddam Hussein. If he had been left in power during a period of worldwide Islamic insurgency, with control over Iraq's oil revenue, and with mountains of conventional ordnance, he would have been in a position to act as a major enabler of dangerous mischief. We forget he had megalomaniacal ambitions of his own. It's worth recalling that the 1991 Gulf War started because, after invading his neighbor Kuwait, Saddam made noises about ultimately extending his reach into Saudi Arabia. Ten years later, the attacks of September 11, 2001, also placed the United States on alert against any previously unthinkable form of aggression issuing from that part of the world. Nothing could be ruled out. That Saddam was not involved in the 9/11 attacks was no guarantee that he would refrain from sponsoring terrorist acts in the future. In fact, the 9/11 attacks might have inspired him to try an extravaganza of his own. In the midst of an international Islamic uprising, the last thing the United States needed in the Middle East was a maverick maniac in a geographically pivotal position. And so the eviction of Saddam became inevitable. The difficulties of the U.S. occupation following his eviction are another issue entirely.

The United States was also concerned strategically that Iran was on its way to becoming a nuclear power whose aggressive tendencies in fomenting Islamic revolution would be greatly enhanced if it acquired nuclear leverage. A U.S. military presence in Iraq would have a tempering influence on Iran's nuclear development program and its messianic impulses. Iran, the world's fourth leading oil producer, was run by a revolutionary Shi'ite mullahocracy. Neighboring Iraq itself happened to have a majority Shi'a population, which was subject to the influence of Iran's mullahs. Iran had suffered tremendously in the decade-long war started by Saddam Hussein in the 1980s and didn't want anything like it to happen again. The establishment of a Shi'a majority government replacing Saddam was in Iran's interest. Ironically, the quickest path to that was via "the Great Satan," America, getting rid of Saddam one way or another for them. The Iranians were disappointed by the lack of follow-through in the 1991 Gulf War. When American ground troops withdrew in 1991, they had fomented a Shi'ite uprising in the Iraqi oil field region around Basra, which was brutally put down by Saddam. The Iranians hoped that the United States might provoke some internal coup d'état against Saddam in the ten years that followed, but it was not to be.

After the 2003 American invasion of Iraq, Iran sought again to manipulate Shi'ite forces within Iraq to take leading roles in the new government. The United States was caught in a predicament. The non-Shi'ite

Iraqis, namely the Sunni, were disposed to side with al-Qaeda-led radical Islam. The deposed Saddam Baathist remnants were themselves Sunni (though they had run a regime patterned more on the Nazis than on any other model). These Sunni and their Baathist cohorts, along with volunteer mujahedeen from other countries, including al-Qaeda fighters, had been leading the insurgency of suicide bombings and sabotage during the U.S. occupation. Therefore the United States was inclined to stack the new Iraqi government with Shi'ites, on the condition that the government they formed was one of secular law, not Islamic law.

That is where matters stand as of this writing, with a provisional government launched in June 2004, led by a Shi'ite prime minister, Iyad Allawi, running what is so far a secular interim government. Three complaints have been lodged against the current American policy. One is that attempting to democratize Iraq is a folly. This may be so. But by "democratize," we mainly mean holding elections so the Iraqi people can choose their own government. It doesn't mean forcing them to adopt a menu of permanent democratic institutions against their will. Besides, what would be the alternative to elections, inasmuch as we are now there and cannot take back the invasion? To simply install a new dictator? Would the other nations of the world look more favorably on that? So, the United States is doing the only other possible thing: attempting to set up a mechanism by which the Iraqi people can elect their own leaders. Will it succeed? Perhaps not. The Iraqis themselves may be too fractious to support an elected government, and the pressure of the insurgency might grow even worse. The effort may be too difficult and costly for the American public to tolerate. Ultimately, the United States might be driven out by a combination of Iraqi resistance and political pressure back home. We certainly can't expect to occupy Iraq forever, or even for a modest fraction of forever.

The second complaint is that the occupation itself was botched. It certainly wasn't the lovefest some expected from a grateful liberated people eager to be democratized. It is true that many things did not go well, from the looting that erupted hours after the United States entered Baghdad, to the sustained terror bombing and assassination campaign that got under way in the spring of 2004, to the chronic failure to restore electric service and water, to the further failure of not getting Iraq's battered and neglected oil infrastructure back in working order—which was the explicit job of contractors Halliburton and its subsidiary, Kellogg Brown, and Root (Vice President Dick Cheney was CEO of Halliburton before the 2000 election). The last was especially vexing, as revenue from Iraq's oil sales was supposed to partially finance the reconstruction of the country. One could legitimately question the foolishness of decommissioning Saddam's army and thus leaving thousands of warriors without income or useful work while giving them an incentive to join an insurgency. Finally, there was the disgraceful and thuggish behavior of American military jailers at Abu Ghraib prison. The occupation may have been a poor performance. Or it may have been nothing more than the ordinary (and legendary) incompetence of the military at war. Or it may even have been a case of the normal vicissitudes of war. What is perhaps more interesting is the American public's increasing intolerance for anything less than guaranteed 100 percent happy and successful outcomes—a result, in my opinion, of watching too many television shows and being too accustomed to happy endings.

The third complaint, this one made by the American antiwar lobby, is that "it's just about the oil." Of course it was about oil. The Iraq invasion was a desperate attempt by the United States to establish political stability in the Middle East, where so much of the industrial world's oil comes from. But members of the antiwar lobby were just as likely to be car-dependent suburbanites as Bush supporters were. At least that was my observation among my fellow middle-aged yuppies in upstate New York. None of them had traded in their giant cars or scaled down their driving habits or moved closer to town or done anything to make their lives less reliant on liberal supplies of Middle Eastern oil. One family in my neighborhood had a sign in their yard that said "War Is Not the Answer"—and had two SUVs parked in the driveway. The American public, including the educated minority, seemed eerily clueless about the connection between their own living arrangements and our problems overseas.

No End of Trouble

As insurgency raged in and around Baghdad through 2004, Iran resumed its work on nuclear development, despite the presence of the U.S. military next door, perhaps in defiance of it like the Iraqi insurgents. Iran's leaders maintain, with some logical justification, that their nation's oil and natural gas will not last forever and it has to make provision for generating electricity when oil becomes scarce—an argument that would apply equally to the United States. It is an additional fact that Iran's oil industry is one of the oldest in the world—first developed by the British around 1900—and that it is indeed well past its own production peak, just as the United States is. Technically speaking, however, a nuclear industry capable of running power plants would also be capable of producing material for nuclear weapons—a fact that leaves Iran in a quandary. It is generally believed that the United States will not tolerate a nuclear Iran, and it is also generally believed that Israel would be even less equivocal about it. It is believed further that Israel, acting either on its own or as a proxy for America, will destroy Iran's nuclear installations before they are ever operational, just as it destroyed Saddam Hussein's Osirak reactor by air in 1981. Policy analysts and media commentators often say that a strike against Iran by the United States or Israel would have the potential of igniting a third world war. One might take the view, however, that World War III has already started and we are well into it. The Bush doctrine of calculated preemption is a clear announcement that the terms of engagement have changed. Now, with George W. Bush reelected, and Iran still labeled as an odious member of the "axis of evil," the chances that Iran will be allowed to go forward with a nuclear program do not seem good. The United States has forces poised not only next door to Iran on both the east and west, in Iraq and Afghanistan, but also to the north in Azerbaijan and to the northeast in Uzbekistan, Tajikistan, and Kyrgyzstan, not to mention substantial naval forces to the south in the Persian Gulf and the Arabian Sea. In short, Iran is surrounded by American military force, within minutes of striking by air. This must make the Iranians extremely uncomfortable. We ought to hope that it doesn't make them crazy. Americans also need to consider how long we can keep all these troops in place and in a state of readiness. The expense alone must be forbidding.

Meanwhile, there is the grinding insurgency next door in Iraq. It is impossible to predict what will happen in Iraq except this: The United States cannot operate its Iraq Middle East police station indefinitely. Nor can the United States remain indefinitely in a place of permanent insurgency. While many foreign affairs experts point out that Iraq is not Vietnam, the fact is that we were forced out of Vietnam precisely because it was in a state of permanent insurgency. If and when we do have to quit Iraq, we will probably be saying good-bye to any claims on the oil of the Middle East, so a lot is at stake and the prospects are rather dark. By then, the Long Emergency will be well under way and the United States itself may be in a state of political turmoil.

What about Afghanistan?

We bombed Afghanistan and invaded and occupied it beginning in October 2001 because the Taliban government allowed its territory to be used as a headquarters for the group that launched the September 11, 2001, airplane attacks against the United States. It may have been a grave mistake that the United States waited nearly a month after the attacks on the World Trade Center and the Pentagon to launch its counterattack, because it gave the leaders of the Taliban and al-Qaeda plenty of time to escape next door into the friendly precincts of sovereign Pakistani territory. Our invasion yielded little besides the occupation of Kabul, the capital. Three years later, most of Afghanistan outside Kabul remains under the control of ethnic warlords, while many Taliban and their al-Qaeda allies have returned to their familiar haunts in the rugged backcountry. For all that, the election that took place in October 2004 was a remarkable first in that nation and Hamid Karzai presides over his tiny government in a very large and unruly territory with something like legitimacy. However honorable he may be, though, the joke is that he is president of Kabul. He is certainly brave, inasmuch as he has already survived more than a couple of assassination attempts. In any case, American soldiers are there to protect him for the time being.

It is a cliché now that Afghanistan is the elephant's graveyard of imperial overreach, a remote, arid mountain terrain inhabited by savage and implacable primitive warriors armed with little more than repeating rifles, a few mortars, and shoulder-launched rockets. Yes, the British were taught a harsh lesson there in the nineteenth century and the mighty Soviet army was driven out by a few bands of pajama-clad mujahedeen in the twentieth century (armed with deadly U.S.-made shoulder-launched Stinger missiles). For the moment our purpose there is to maintain a forward military base on the west side of Iran, as well as to influence the behavior of Pakistan to the east. Pakistan, with its ten to twenty known nuclear bombs, its military and intelligence services riddled with Islamic fundamentalists, its chaotic economy, and its precarious central government, is one of the most dangerous and unstable nations on earth and needs to be closely monitored. Whether the United States can accomplish that from Afghanistan is far from certain, but that is why we remain there.

There was, briefly, another reason. Once America had established a foothold in Afghanistan, plus military bases in Azerbaijan, Uzbekistan, Kyrgyzstan, and Tajikistan, a notion developed among U.S. government officials and major oil company executives that a pipeline could be routed through Afghanistan and Pakistan from oil-producing former Soviet republics to the Indian Ocean. This was viewed as a neat way to bypass the chokehold of Saudi Arabia on U.S. oil imports. But two developments killed that idea once we became familiar with the terrain and its inhabitants: first, the dawning recognition of our absolute inability to defend such a pipeline against casual sabotage, and second, the fact that the oil reserves in the former Soviet republics proved to be both much lower in quantity than originally estimated, as well as lower in quality, being a high-sulfur "sour" crude that it is much more difficult and expensive to refine than the "sweet" crudes of Arabia and elsewhere.

For the time being, however, we remain in Afghanistan for the other strategic reasons, and to prevent it being used again as a base for militant Islamic terrorist training. How long we remain there depends on similar cost contingencies as our situation in Iraq, with one additional qualification. Afghanistan is virtually next door to China (they actually share a tiny sliver of border). Sometime in the not-too-distant future, an aggressive China might have more to say about who controls the territory in Central Asia. When the time comes, it is unlikely that we will want to engage in a land war in that part of the world with a nation that has the world's largest infantry.

Europe and Russia

Europe and Russia face many of the same problems as the United States does in the Long Emergency that will attend the end of the cheap-oil era. In some respects they are better prepared for what is coming, and in other ways less fortunate. They will all be subject to hardships arising out of oil and gas scarcities, but their geopolitical situations have some differences during these "rollover" years of peak production. Geopolitically, Great Britain is a special case among the other European nations, as it was a major partner in the Iraq war and thus linked much more closely to the United States. That said, and given Britain's large and sometimes overtly belligerent Muslim population, it is something of a miracle—or perhaps a tribute to the British intelligence services—that a major terrorist incident has not occurred there since the March 2003 Iraq invasion.

To some extent France, Germany, Great Britain, and Russia have enjoyed a state of denial equal to that of America in these years leading up to the permanent energy crisis, but their denial is focused differently. America has been able to pretend that its energy-intensive suburban lifestyle is "not negotiable." France and Germany have been able to pretend that they are immune to the geopolitical struggle with Islam and steered clear so far of overt hostile engagement with al-Qaeda and its sympathizers, despite the fact that they have substantial Muslim populations that display often belligerent disaffection for their hosts. Spain was bloodied in the 2004 Madrid train bombings for assisting minimally in the U.S. occupation of Iraq, and quickly withdrew her token forces afterward. Italy had an even smaller symbolic force of military police in Iraq. Several Italian MPs were killed by insurgent assassins in early 2004, and Italy was warned via the Internet that it would suffer homeland attacks, which have not yet materialized as of this writing in late 2004. France and Germany have both managed to stay outside the fray and appease the international forces of radical Islam—including substantial hostile Muslim immigrant populations in their own countries.

Russia has been engaged even longer than America in a bitter and costly struggle with Islamic insurgents, going back to the 1980–89 war in Afghanistan before the Soviet Union broke up. If anything, Russia's actions against Islamic insurgents have been more brutal than America's, and undertaken with more conviction. Russia's southern frontier is composed of Islamic nations, former Soviet colonies with some bad feelings about it. Russia has been attacked repeatedly in its homeland over the past decade, and the September 2004 massacre of school children in Beslan was arguably a moral insult on a scale equal to the 9/11 attacks in America. Hence, Russia has hardly been in a position to object to America's use of force against militant Islamic fundamentalism, even though it is deeply uncomfortable with U.S. military bases in the former Soviet republics.

Russia is currently the world's second largest oil producer, a circumstance that has given it a false sense of security. Russia's particular brand of denial issues from the fact that it passed its own oil reserve peak in 1986 and in the not-distant future will find itself in a predicament similar to America's: facing the loss of energy self-sufficiency. Russian nuclear power facilities are unspeakably decrepit. Russia also has a roster of other obvious problems related to its difficult transition out of the Potemkin communist economy. It still has a long way to go to establish a society of law, in particular a framework of property and contract law that allows business to be conducted on some basis other than gangsterism. In addition, Russia already faces the kinds of problems that are a preview of the Long Emergency: falling standards of living among the masses, lowered life expectancy, deindustrialization, and daunting environmental destruction as a legacy of the communists.

France, Germany, and Russia all had substantial financial stakes in Saddam Hussein's Iraq, and were reluctant to kiss off billions of euros and rubles by assisting in toppling him. Once Saddam's overthrow was established fact, their contracts were effectively worthless. France and Germany remain terrified of provoking their own Muslim populations, and they have relatively fresh memories of their own episodes of terrorism in the late twentieth century—the 1995 Paris subway bombings by Algerian Islamic maniacs, the 1972 Munich Olympics massacre by the Black September group. Despite its gingerly behavior on the international scene, France has begun to get tough with its large Muslim population. A new law banning religious costumes in schools and other public places went into effect in the fall of 2004 without any acts of civil disorder.

France was prescient enough over the past three decades to construct a network of nuclear generating plants that supply roughly 80 percent of the country's electric power—far more than any other nation. France also employed a uniform design standard for all its reactors that has made for an exceptionally safe and well-run system. Alone among the nations of Europe, France is planning a new generation of nuclear power installations. Germany and Belgium, for example, are looking to shut down their existing nuclear power facilities. In any case, all the nations of Europe, including France, will have problems when oil and natural gas grow scarce. France may differ somewhat in being able to keep the lights on longer.

Great Britain is in a particular state of denial and jeopardy over its energy prospects. The bonanza of the North Sea oil and gas fields induced a dangerous sense of euphoria there. For two decades Great Britain was a net oil exporter. Its newfound oil riches supercharged the economy. Now that the North Sea fields have passed their peak and are depleting at about 5 percent a year, Britain faces a bleak energy future. To make matters worse, the North Sea bonanza provoked a spate of suburban development across England that will prove to be an additional burden. All of Western Europe for the foreseeable future will have to rely on natural gas imports from Russia—a dangerous dependence in any case—and Great Britain will find itself near the end of a long pipeline delivery system, ahead only of Ireland, which faces an even more dire energy future. Any time Russia sneezes, Great Britain will have to worry about catching a cold. Germany is not much better off in that respect, made worse by its status as Russia's historical enemy. Germany has next to zero oil and gas resources and not much of a Plan B, despite a big push to develop wind farms on the North Sea.

That said, the nations of Europe enjoy some advantages over the United States in facing the Long Emergency. Although all European countries have some suburban development, it is nowhere comparable to the complete fiasco of American suburbia, and they did not trash their towns and cities in the process, as America did. The quality of compact urbanism, the scale of it, and its integral nature, even in large European cities, is much more sustainable than anything found in America. Most European cities, including the big ones, are still composed primarily of buildings under seven stories at their centers. Because of their high density and the fact that the middle class and even the rich inhabit their centers, even small European cities have high levels of amenity and culture. If there were a major supply interruption in oil, most Europeans would still be able to get to work and carry on the business of their societies. Public transit is still excellent in most of Europe at all scales, from city subways, streetcars, and buses to major passenger rail—though England, in its twenty-year North Sea oil mania, allowed its train system to go to seed. As in America, the megacities of Europe will suffer from their gigantism in the Long Emergency and unquestionably undergo significant contraction. But their centers may hold; in America, most cities no longer have any center.

Finally, the Europeans did not allow local agriculture to be overwhelmed by corporate gigantism and industrial totalism. There is still a clear distinction between urban and rural life in Europe, and virtually all cities and towns are surrounded by active agricultural hinterlands. The degree of local value-added activity associated with European agriculture—winemaking, cheese and olive oil production, and the like—remains high and retains high levels of craft quality. The European Union has led to some decay of the special protections and subsidies enjoyed by farmers, but local agriculture is still an ongoing enterprise there. The cultural memory has not been erased, as it has in the United States. Since local food production will be a crucial issue in the Long Emergency, the Europeans have better prospects of being able to feed themselves. Only one other factor looms ominously offstage at this point: the effects of climate change. Will Europe heat up, or will hydrothermal changes in the Gulf Stream plunge it into icy cold? Something is happening and we do not know the answer yet.

Global Turbulence

For the most part, Europe has been able to stand back from the growing turmoil in the Middle East and let America do all the dirty work of attempting to salvage security and order there—during which time Europeans have benefited, at least, from still relatively stable oil markets. This can change at any moment, of course, and certainly will sooner or later. Though the charge is constantly made that the United States is there only for the oil, it ought to be pointed out that to date we have at least been willing and able to pay whatever the going rate is. It's not as if we've been stealing it—though radical Islam might say that any price for us is too cheap. In any case, political conditions around the world are primed only to change more radically as the peak oil rollover occurs and real scarcity asserts itself. In trying to predict what will happen, we confront a flow of events and possibilities that are nonlinear, chaotic, emergent, and self-organizing. But I'll venture to imagine the following, with the understanding that it is just conjecture, educated guesswork:

Before much longer, Europe may be forced to join the fray against militant Islamic fundamentalism, especially if a revolutionary regime change were to occur in Saudi Arabia, and if that resulted in a restriction of exports by policy or by sabotage. If the pipeline from the Persian Gulf to the Red Sea is severed, Europe will be in big trouble, as 3.8 million barrels a day pass through Suez.[5] If European engagement with militant Islam led to increased terrorist incidents in Europe, things could get ugly. Were terrorist acts to resume in France, that country's established anti-Semitism might be redirected against its Muslim population, leading to deportations, restrictions of civil liberties, or worse. Germany has not flexed its military muscles in half a century. We might be surprised by what it brings to the battle—perhaps a desire by Germany to have a larger say in the Middle East specifically and in an increasingly competitive international arena generally. It has been a long time since we had to contend with Germany as a world power. It is not inconceivable that the Middle East might come under the hegemony of NATO, or some combination of Euro-American military power, for a period of time. Even Russia might be induced to participate, given its problems with Islamic insurgency.

America with or without Europe is likely to confront an energy-starved China in the next decade or two. There is no telling how desperate or politically psychotic China might become, or how aggressive in its reach. A land war against China over Central Asia is pretty much unthinkable for either America or Europe—especially with Islam rampant in the foreground. But if successful in gaining control over some former Soviet republics, what is to stop China from continuing into Iran, Iraq, and even Arabia? Russia's nuclear arsenal perhaps. Perhaps India, another nuclear

5. Energy Information Administration (EIA), the statistical agency of the U.S. Department of Energy (DOE).

power, with a population set to surpass China's in the early twenty-first century, might act as a check on Chinese adventurism. A military contest over oil could eventually inflame a theater of war stretching from the Middle East to Southeast Asia, and it could leave the oil production infrastructure of many countries shattered in the process. Such a conflict might be the Last World War.

Whether the nations of the world fall into war over the remaining oil remains to be seen. What is certain is that we are entering a new period of world history, the uncharted territory of a post-oil world. We will be in it long before the middle of the twenty-first century. Eventually, all the nations will have to contend with the problems of the Long Emergency: the end of industrial growth, falling standards of living, economic desperation, declining food production, and domestic political strife. A point will be reached when the great powers of the world no longer have the means to project their power any distance. Even nuclear weapons may become inoperable, considering how much their careful maintenance depends on other technological systems linked to our fossil fuel economy.

Before long, all nations will retreat back into themselves either in autarky or anarchy. Many of them—including possibly even the United States—will probably follow the example of the Soviet Union and fragment into smaller autonomous units, as life becomes intensely local everywhere. I have not mentioned South America so far in this chapter, for the simple reason that I think it will remain off on the geopolitical margins into the Long Emergency. This does not mean it will be a safer and happier place than the rest of the world. South American countries will have to contend with exactly the same problems of energy scarcity, falling food production, and all the rest. Given the already high level of violent anarchy in many regions of South America, we can expect at least a continuation of armed conflict and disorder. South American nations will not, however, be in much of a position to project their power into the Eastern Hemisphere. Mexico, Colombia, and Venezuela may find themselves combatants in their own oil wars, though they are all well past peak.

Australia and New Zealand may fall victim to desperate Chinese adventuring, or to anarchy emanating from Southeast Asia. Or perhaps they will be left alone. In any case, both will be starved for fossil fuels. Africa, too, has been left to the margins in this discussion. While it contains some major oil producers, the continent is already a poster child for the kind of hardship and chaos that will become common elsewhere. The oil-producing nations there may easily become too disorderly to even support the continued exploitation of their oil fields.

In the Long Emergency the world would become a larger place again. Globalism, as a set of economic relations, will fizzle out. The 12,000-mile supply lines from the factories of Asia to the Wal-Marts of Pennsylvania will be a thing of the past. Commercial sea lanes might become indefensible. In fact, the coastlines of all nations may become prey to a new species of stateless freebooting raiders, as the area around the Molucca Sea is now. I believe the Pacific coast of North America will be especially vulnerable to raids emanating from the disintegrating nations of Asia. Air transport could become a rarity or a prerogative of only small and diminishing elites. Finally, the international oil trade itself would become so chaotic and unmanageable that no region on earth would be able to rely on distant energy supplies any longer. Nations, and even more likely regions within nations, would have to fall back on their own resources, and sink or swim.

FOUR

BEYOND OIL:

Why Alternative Fuels Won't Rescue Us

Based on everything we know right now, no combination of so-called alternative fuels or energy procedures will allow us to maintain daily life in the United States the way we have been accustomed to running it under the regime of oil. No combination of alternative fuels will even permit us to operate a substantial fraction of the systems we currently run—in everything from food production and manufacturing to electric power generation, to skyscraper cities, to the ordinary business of running a household by making multiple car trips per day, to the operation of giant centralized schools with their fleets of yellow buses. We are in trouble.

The known alternatives to conventional oil that I will discuss in this chapter include natural gas, coal and tar sands, shale oils, ethanol, nuclear fission, solar, wind, water, tidal power, and methane hydrates. We will certainly use many of these things, and the various systems they entail, as much as we can, but they will not make up for the depletion of our oil supply. To some degree, all of the non-fossil fuel energy sources actually depend on an underlying fossil fuel economy. You can't manufacture metal wind turbines using wind energy technology. You can't make lead-acid storage batteries for solar electric systems using any known solar energy systems.

The pseudo-fuel hydrogen will be considered in its own special category, as the popular hopes about it are based on higher orders of unreality. The so-called "hydrogen economy" centered around hydrogen-powered cars, as promised by President Bush in his 2003 State of the Union message, is at this point a fantasy, and an especially dangerous one insofar as it promotes complacency about the predicament we face. If there is ever going to be a hydrogen economy, then we are not going to segue seamlessly into it when the fossil fuel economy begins to wobble. At best, the world is going to suffer an interval of economic chaos and social stress between the end of the fossil fuel age and whatever comes next. The question is how long this interval will last: ten years, a hundred years, a thousand years, or forever.

The belief that "market economics" will automatically deliver a replacement for fossil fuels is a type of magical thinking like that of the cargo cults of the South Pacific.[1]

This age-old tendency of humans to believe in magical deliverance and to wish for happy outcomes has been aggravated by the very technological triumphs that the oil age brought into existence. Technology itself has become a kind of supernatural force, one that has demonstrably delivered all kinds of miracles within the memory of many people now living—everything from airplane travel to moving pictures to heart transplants. There's no question that technology has prolonged life spans, relieved misery, and made everyday life luxurious for a substantial lucky minority. (The diminishing returns and unintended consequences of technology are important topics that will be explored later in Chapter Six.) A hopeful public, including leaders in business and politics, views the growing problem of oil depletion as a very straightforward engineering problem of exactly the kind that technology and human ingenuity have so successfully solved before, and it therefore seems reasonable to assume

1. In the eighteenth century, Europeans first appeared in that part of the world, bringing all sorts of wondrous cargo with them in awesome great sailing ships—telescopes, cannons, felt hats, pocketknives, metal cookpots, you name it—and astounding the natives. When the Europeans sailed away, as they did periodically for long periods of time, the islanders would build effigies of the ships out of whatever plant materials they had at hand in an attempt to lure back the great ships and all the fabulous stuff they had brought with them. This behavior was seen again after World War II. The Pacific campaign had drawn huge numbers of men and materials to the South Sea Islands, often in airplanes. When the war ended, the bereft islanders placed rattan effigies of B-23s on mountaintops, hoping to lure back the airplanes and all the wonders they'd brought. The Islanders would lay out pretend runways, light fires along the sides, make a wooden hut for a man to sit in with wooden headphones on and palm fronds sticking out like antennas, and they'd wait for the airplanes to land. "They're doing everything right. The form is perfect. It looks exactly the way it looked before. But it doesn't work. No airplanes land." Physicist Richard Feynman described this phenomenon humorously in a 1974 commencement speech at CalTech. Cargo cults are also discussed in detail in Marvin Harris's excellent book *Cows, Pigs, Wars and Witches: The Riddles of Culture*, New York: Random House, 1974.

that the combination will prevail again. There are, however, several defects in this belief.

One is that we tend to confuse and conflate energy and technology. They go hand in hand but they are not the same thing. The oil endowment was an extraordinary and singular occurrence of geology, allowing us to use the stored energy of millions of years of sunlight. Once it's gone it will be gone forever. Technology is just the hardware and programming for running that fuel, but not the fuel itself. And technology is still bound to the laws of physics and thermodynamics, which both say you can't get something for nothing, and there is no such thing as perpetual motion. All of this is to say that much of our existing technology simply won't work without petroleum, and without the petroleum "platform" to work off, we may lack the tools to get beyond the current level of fossil-fuel based technology. Another way of putting it is that we have an extremely narrow window of opportunity to make that happen. In the meantime, here are the problems with the various alternative fuels, based on what we know now.

Natural Gas

For the sake of this discussion, by natural gas I mean methane. Of the various natural gases that come out of the ground, methane (CH_4), the lightest of them, makes up 75 percent of commercial product used in industry, electrical power generation, and home heating. The others, propane, butane, and so forth, are separated in processing and lend themselves more easily to liquefaction because they are denser and heavier.

Natural gas is colorless and odorless. A trace amount of dimethyl sulfide is added to commercial gas to give it a detectable nasty odor so people will know if it's leaking (and be alarmed). It is explosive when mixed with air in concentrations of 5 to 15 percent. Natural gas is created the same way as oil but in geological conditions of greater heat and pressure—as the folded rock strata are thrust deep below the oil "window" by tectonic forces—and gas deposits are commonly associated with oil fields. Roughly one-third of the total energy used in the United States is derived from natural gas.

Natural gas is a wonderful fuel. It comes out of the ground easily, under its own pressure, without the energy "input" of pumping. (It can also be distilled out of coal, but the cost of mining the coal, as well as the energy used in distillation, adds to the price.) It is a "clean" fuel. It produces little to no particulate matter when burned, but it does give off carbon dioxide, the major "greenhouse" gas. Natural gas is transported easily at air temperature through pipeline networks all over North America connecting the wells and storage facilities to the end users. It is not as versatile as gasoline, but it does a lot of tasks beautifully. Gas is the feedstock—the raw material—for a wide array of chemicals, pharmaceuticals, and plastics. Ninety-five percent of the nitrogenous fertilizers used in America are made out of natural gas, and so it has become indispensable to U.S. agriculture.

In the early twentieth century, natural gas was so plentiful that it was regarded as an annoying waste by-product of the oil industry and was routinely "flared off" or burned at the wellhead. After World War II, the construction of a comprehensive national pipeline network made gas a profitable commodity. U.S. oil production peaked in 1970, but natural gas production passed its own peak a short time later in 1973 at 22.90 trillion cubic feet (Tcf) and has been declining ever since. Higher-efficiency furnaces, fluctuating demand, and flip-flopping regulations helped to obscure this fundamental fact until the twenty-first century.

Ironically perhaps, the OPEC oil embargo of 1973 prompted many homeowners to switch from oil to natural gas furnaces the very year gas production peaked, though the peak would be seen only in retrospect. Gas was clean, cheap, and produced right here in the United States. One's comfort in winter was not at the mercy of foreigners. The traumatic OPEC embargo also promoted the general idea of energy conservation, which led to more efficient gas-burning technology. By 1978, however, a marked decline in gas production was evident. The Carter administration figured depletion into its energy policy, making it illegal to use natural gas or oil as the fuel for any new plants built to generate electricity. Under the Carter policy, coal and nuclear power plants were encouraged to meet new demand. Then, in March 1979, the Three Mile Island nuclear plant near Harrisburg, Pennsylvania, suffered a partial meltdown, which put any growth in the U.S. nuclear industry on indefinite hold. The environmental legislation of the 1970s made coal increasingly problematic, too, as it was implicated in acid rain.

Meanwhile, by the mid-1980s consumption of natural gas had dropped by 24 percent from the early 1970s levels. The natural gas producers began to go bankrupt. To rescue the industry, the Reagan administration reversed the Carter regulations. Instead of forbidding natural gas in power plants, the regulators now encouraged it. The April 1986 meltdown of Chernobyl in the Ukraine was much worse than Three Mile Island had been. The Chernobyl horror story virtually killed any prospects for ramping up the American nuclear industry, since the NIMBY ("not in my backyard") reaction now would be insurmountable. The United States was in a quandary. Both natural gas and oil were in depletion domestically. All the major hydroelectric sites were already in use. Coal was dirty. Nuclear fuel was politically untouchable. We were importing nearly half our liquid petroleum by then and didn't want to go through another foreign energy blackmail crisis. Though the United States was producing less gas than it had years earlier, we were using less, too, and future capacity could come from Canada, our friendly neighbor. By default, then, gas became the least unpleasant choice for meeting future electric power demand. More than 275 North American gas-fired electrical generation plants were planned to begin operations through 2006, up from 158 operating in 2000, which would increase gas consumption by more than 8.5 trillion cubic feet a year.

By the year 2000, despite improvements in drilling technology, favorable tax credits, and an intense exploration effort in the Gulf of Mexico, U.S. natural gas production was still 10 percent less than production in 1973. The gap between consumption and production was met by increasing imports of gas by pipeline from Canada and small volumes of liquefied natural gas (LNG) imported by ship. Despite the fact that it is a major oil producer, Mexico has become a net importer of natural gas. Ironically, the North American Free Trade Agreement (NAFTA) compels the United States to sell Texas gas to Mexico, which the United States then must replace with imported Canadian gas. Canada, also past its natural gas peak, is in turn required under the NAFTA treaties to sell gas to the United States at market prices.[2]

2. While technically in depletion—that is, its production steadily declining—Canada still has enough gas to export to the United States and meet domestic needs, but the time may not be far off when it cannot do both.

In 1999, the National Petroleum Council predicted that the growth of gas supplies would be adequate to meet demand increases of 36 percent by 2010. The organization's forecast proved wrong beyond its experts' wildest dreams. American natural gas production is now firmly declining at 5 percent a year, despite frantic new drilling, with the potential of much steeper declines coming. The 167 giant gas wells that made up 14.5 percent of the 2001 supply produced only 3 percent of total supply by 2003, a fall of 82 percent.

When gas wells go into depletion, the stuff just stops coming out. Unlike oil wells, which go from gushing at high pressure to a moderate flow for a long period of time, and finally to a slow dribble (often mixed with water), all in a predictable way, gas well production falls off a cliff, often with little warning. This is exactly what has been happening to American wells. Depletion rates of individual fields have been climbing sharply in the past decade. The only thing keeping gas production even close to par since the year 2000 has been new drilling, but the newest fields have been depleting with especially alarming speed, many in less than a year of operation.[3] This may be partly attributed to improved drilling technology, and partly to the fact that the newer fields are much smaller than the old ones. Discovery of new gas fields in the United States, meanwhile, is declining steeply, just as discovery of new oil fields has been decreasing steeply worldwide. Depletion of onshore U.S. gas fields was so uniform that exploration on dry land has virtually halted. The only new gas to be found within U.S. sovereign territory lies underwater, mainly around the Gulf of Mexico.

U.S. gas supplies were so low in March and April 2003, after an extremely harsh winter, that officials faced potential triage operations to sequentially shut down delivery to end users in a rational way to preserve life and property. This meant cutting off supplies to manufacturers first,

3. David Pursell, U.S. Department of Energy, Energy Information Agency (Report SR/OIAF/2000-04): "Accelerated Depletion: Impacts on Domestic Oil and Natural Gas, Appendix G." "Twenty-three months after reaching peak production in January 1997, the average production from natural gas wells that began producing in 1996 was 69 percent lower than it had been at its peak." Pursell cites the increase in the rate of decline in natural gas production per well from less than 20 percent per year in 1970 and 1971 to 49 percent per year for wells completed in 1996.

then to electric power plants, and last to home heating customers. The theory behind this was that people would be better off sitting in a warm house in the dark than helplessly watching their frozen pipes burst with the lights on. It never came to that, but it came close enough to scare those in business and government who were watching in horror.

There were other interesting and scary implications of a natural gas shortage: for example, fears that stored reserves of gas would fall so low that the pressure would drop dangerously low in pipeline networks. If the pipeline pressure to a given city neighborhood or town fell too low, or if it fluctuated, pilot lights would go out in home furnaces. Most people have no idea how to operate their own home furnaces. Armies of technicians would have to be sent out to restart them. The cost of mounting such an operation would be staggering to an electric utility company. What about the houses that for some reason were overlooked? Once pipeline service was restored to a locality, gas flowing from water heaters and furnaces left in the "on" position at the time of pressure drop could cause explosions when the gas started flowing again. The repressurizing of pipelines would be a difficult and costly operation in itself, even after stored reserves were replenished.

As the situation developed in the late winter of 2003, the price of natural gas doubled from roughly $3 per thousand cubic feet to about $6. This led to another consequence of the 2003 quasi-crisis: Many chemical manufacturing companies decided to move operations to other countries, including half the U.S. factories that make fertilizer. It was obvious to them that the natural gas situation in America was not going to improve, and that the long-term picture looked dismal for industries that used gas as the raw material, the "feedstock," for their products. So they began to move to Asia and the Middle East.

What should have sent up a red flag for political leaders and the media has been regarded as just another ho-hum phase of the globalization process. Whatever one thinks of industrial farming based on fossil fuel "inputs"—and I will later discuss at length America's desperate need to reform agriculture—the fact is that this is how we currently produce the bulk of our food (and food for many people in other nations), and to lose control of the basic means of production before we are ready to change could have catastrophic repercussions.

The bottom line is that the United States—indeed, North America, faces a chronic and accelerating natural gas shortage that sooner or later will be described as a crisis. Canada faces compound dilemmas. It already exports two-thirds of its declining gas production to the United States. There is enormous political and economic pressure for Canada to exploit its vast Alberta tar sands, which are said to contain the equivalent of 200 billion barrels of oil (equal to 20 percent of the remaining conventional oil). But processing tar sands into oil is no easy or cheap matter. The tar doesn't come out of the ground the way oil does. Getting it is more like an open pit mining operation, and once dug up the gunky material has to be "washed" with tremendous volumes of superheated water before it can be taken to the refining stage. The process also generates vast quantities of polluted groundwater. Both the mining and the washing require huge amounts of energy, and it has been proposed that any commercial exploitation of the Alberta tar sands would take 20 percent of Canada's total natural gas production. In the long run, it might not be worth expending the energy from gas to get the energy from the tar sands. If oil from the tar sands themselves were used to process more tar sands, the return would be three barrels of oil for every two consumed. The case for oil shale is similar. The cost of separating the oil from the rock matrix at the scale necessary to justify the operation would hardly make the operation economically worthwhile.

This is the classic problem of energy economics: energy returned over energy invested (ERoEI). It applies in one way or another to all categories of fuel and every procedure for getting and using them, and it comes back to a basic law both of physics and metaphysics: You can't get something for nothing. In the early days of conventional oil in Texas, the ERoEI formula was very favorable, around twenty to one. The oil was found close to the surface on dry land in temperate places easy to work in, and it gushed out of the ground under its own pressure. Eventually, when the pressure equalized, it had to be pumped out of the ground, and the cost of pumping lowered the ERoEI somewhat. As producers have moved year after year to extract oil from ever deeper wells in harsher and less accessible places, using more advanced (and expensive) drilling methods, the equation has become far less favorable.

Recovering oil from offshore platforms in the cold and stormy North Sea, for example, is more expensive than drilling through flat-land Texas, though still economically worthwhile. Ultimately, though, the point arrives at which it may still be theoretically possible to get oil out of the earth (and refined and distributed) using marginally less energy than the resource provides, but it may become economically irrational for large corporations to bother doing it, and if they don't do it, who will? Global oil corporations enjoy their economies of scale because of large-scale profit. They are gigantic evolved organisms that came into being for a specific purpose within a specific economic ecology. If some major element of that ecology were to change—e.g., the basic cost/profit equation—the organisms would become extinct, even while there may be substantial deposits of oil, gas, and tar left in the world. Going a bit further, the fundamental equations that support all gigantic global economic organisms, from oil companies to Wal-Mart to nation-states, may no longer obtain, and human life would have to reorganize its activities on a different basis. Also, once these complex systems and their subsystems halt their operations, restarting them may range from difficult to impossible—the Humpty Dumpty syndrome. (I get a little ahead of myself here, but I will explore more fully the issues of systems failure in Chapter Six.)

To return to the issue of ERoEI, future natural gas production in North America may reach a condition of economic nonfeasibility. Depletion is on track to accelerate steeply. No amount of furious and expensive drilling in ever-smaller fields will keep up with the demand. No one will run as fast as they can to fall further behind indefinitely. In the natural order of things, rising prices due to scarcity would initiate "demand destruction." But because there are no substitute energy resources capable of doing the work that gas now does, in precisely the way it does it, such demand destruction would translate into standard of living destruction for the American public. For instance, replacing natural gas home heating with electric heating would probably leave a lot of American families shivering and broke. The political implications are obvious.

The current proposals for enlarging the gas supply are transparently inadequate. A pipeline to the McKenzie Delta region in northern Canada would be costly (about $10 billion), would take years to complete, and might not produce enough gas to make much difference anyway. Ditto the Arctic National Wildlife Refuge (ANWR) in Alaska.

The essential and central fact is that you get the gas that exists on the continent you inhabit and otherwise you have a big problem. Natural gas is distributed around North America by an extensive pipeline system. Small pumps keep it moving at the energy cost of about 0.03 percent of the gas per one hundred miles. This occurs at air temperature. To get natural gas from overseas, it must be liquefied and shipped in special tankers as a highly pressurized supercold liquid. All this requires substantial additional expense. The liquefied natural gas (LNG) is then unloaded into special port facilities at the destination countries, regasified, and distributed into pipelines. The capital costs are so high that this process is economical only under long-term contracts, lasting up to twenty years—and the prospect for long-term international political stability grows worse daily as nations contest over oil and gas supplies. The greatest gas supplies exist in exactly those places—the Middle East and Asia—that are most politically unstable, which is to say with exactly the wrong parties for making credible long-term contracts.

Currently, less than 2 percent of the gas used in the United States consists of LNG imports. In the meantime, the United States is woefully lacking in the port infrastructure for receiving LNG, and the global energy corporations have a very small number of the expensive pressurized tanker ships needed to transport LNG. The U.S. Department of Energy has proposed that at least a dozen LNG receiving terminals be built to avoid a serious supply bottleneck, but events are moving much more swiftly than the federal energy bureaucracy. The United States doesn't have a decade to solve this problem.

It's worth considering that highly explosive LNG tanker ships would make excellent targets for terrorist attacks, and even under normal operating circumstances, liquefied gas transport is much more hazardous than oil. Port facilities and terminals are equally vulnerable to attack or sabotage. Therefore, the political problems in siting LNG terminals are formidable. An LNG terminal is classic "LULU" (locally undesirable land use) certain to provoke a NIMBY response. Finally, even in the unlikely case that a massive system of LNG terminals and tankers could be constructed expeditiously, it is reasonable to doubt that the American public could afford the greatly increased cost of relying on imported LNG to heat their homes and generate electricity. The bottom line: As energy expert Matthew Simmons has said repeatedly, "America has no Plan B."

The Hydrogen Economy

The widespread belief that hydrogen is going to save technological societies from the fast-approaching oil and gas reckoning is probably a good index of how delusional our oil-addicted society has become. The idea is enticing because the only by-product of burning hydrogen is water vapor, and that would seem to obviate most of the world's global warming and air pollution worries. And hydrogen is also a superabundant chemical element. It would be nice, neat, and simple if all the powered infrastructure and equipment of our society could simply be switched to hydrogen, but it's not going to happen. A few things may run on hydrogen, but not America's automobile and trucking fleets. In the long run hydrogen will not replace our lost oil and gas endowments.

Proposals for switching from an oil and gas to a hydrogen economy are generally associated with the fuel cell technology. A single fuel cell is basically a piece of plastic between a couple of carbon plates that are sandwiched between two end plates acting as electrodes. These plates have channels that distribute the fuel and oxygen. They are modular and can be stacked to produce different amounts of power. Fuel cells can operate at efficiencies two to three times that of the internal combustion engine, and require no moving parts. In a kind of reverse electrolysis, hydrogen introduced through a catalytic metal membrane combines with oxygen to produce water vapor and an electric current, which then does useful work. In a fuel-cell car, for example, electricity from the fuel cell would power an electric motor and make the car go. However, due to the cost of making pure hydrogen, most current schemes for mass-market fuel cells propose using natural gas or methanol as the fuel, and that would produce carbon dioxide just like any tailpipe.

Fuel cells have been around a long time. Sir William Robert Grove demonstrated the process in 1839. In the late 1950s, NASA began to build a compact fuel cell electricity generator for use on space missions. Cost was not a constraint. The fuel cells and hydrogen to run them weighed much less than batteries, an important consideration when firing loads into space on rockets. Later, in manned spacecraft, the astronauts could also drink the water that the fuel cells produced.

There is no question that fuel cells exist and that they work. But huge and confounding questions arise over the economics of hydrogen. The problem is that hydrogen is not exactly a fuel. It's more accurately a "carrier" of energy than a fuel. It takes more energy to manufacture hydrogen than the hydrogen itself produces. So at this time hydrogen production depends on the other known energy sources that are all problematic for one reason or another—namely, oil, natural gas, coal, nuclear, hydro, solar, biomass, wind. To some extent, the term "hydrogen economy" is a disguise for "nuclear economy," because nuclear energy may be the advanced societies' only realistic resort where large-scale electric generation is concerned, and the subtext is that an expanded and updated array of nuclear plants could produce large amounts of hydrogen economically. But I will get back to the question of nuclear energy itself later in this chapter.

Of course hydrogen is produced commercially now and has many industrial and chemical uses. But compared with the oil we burn, the amount of hydrogen used by industry is minuscule. Using hydrogen as an industrial catalyst or chemical ingredient is one thing, but it is quite another to propose burning hydrogen as an energy commodity. Where running hundreds of millions of cars is concerned, hydrogen just doesn't scale, as the engineers say. The amount of hydrogen needed to power the U.S. car fleet would be orders of magnitude greater and a net energy loser. We'd get less energy out of the hydrogen than we would put in to create the hydrogen, so what would be the point? The "hydrogen economy" fantasy also does not address the issue of replacing oil and gas to heat tens of millions of houses and other buildings.

Hydrogen composes about 73 percent of all the matter in the universe—at least in our neighborhood of the universe. But it is not found naturally in a free state near the planet Earth. Around here it is always bound to other elements in chemical compounds. Water, H_2O, is the most common: two hydrogen atoms bound to one of oxygen. Hydrocarbons such as oil and natural gas (methane) are naturally occurring hydrogen compounds that can burn and release energy.

Why not just try to synthesize oil and gas from abundant hydrogen and carbon? Because the procedures for first freeing the hydrogen and then combining it with carbon would also take more energy than the resulting compound would produce. (Synthesizing gasoline from coal is a different matter, as that is a case of refining one hydrocarbon to obtain another, and it is still very costly.) Natural hydrocarbons represent millennia of stored solar power collected by plants and distilled by geologic accident. The flare given off by igniting an ounce of charcoal starter lasts a few seconds, but the energy was derived from, say, a prehistoric tree fern absorbing sunshine for nine years. (The hundred-year duration of the oil-powered civilization is nothing compared to geologic time.) Oil and gas are nonrenewable and limited in supply. We can't fabricate them artificially from free elemental hydrogen and carbon without energy inputs that would exceed the fuel value of the hydrocarbons made. Such is the dilemma. As far as pollution is concerned, the procedures used to synthesize methane (CH_4) from coal and methanol (CH_4OH) from oil and biomass produce more carbon dioxide than if the hydrocarbon precursors were burned themselves, so there's no benefit from an air quality standpoint.

Water, on the other hand, is not combustible. To free the combustible hydrogen atoms from the oxygen atoms in water requires a lot of energy. Electrolysis is one method. You run an electric current through a vessel containing water and capture the "cracked" gases separately, based on the fact that hydrogen is much lighter than oxygen (it consists of fewer protons, neutrons, and electrons) and rises to the top of the vessel. The electricity for the process itself has to be generated by using some other fuel. Another way of getting hydrogen is to superheat water to "wash" natural gas at very high pressure, which "strips" off the hydrogen atoms. Of course, this assumes that there are plentiful supplies of natural gas to use as the feedstock, which may be assuming too much. And it would take plenty of energy to superheat the water. These processes for "freeing" hydrogen are always associated with net energy loss. The ERoEI median ratio of these various processes is about one to 1.4. That is, you get one unit of energy from every 1.4 units invested. All of them are net losers. Compare this to the twenty-to-one ERoEI on Texas oil in the 1930s and it's easy to see why oil was such a boon.

There are many additional problems with hydrogen as a replacement for the hydrocarbon fuels we are currently using to run industrial civilization. These have to do with storage and transport. The extremely low density of hydrogen, given its low atomic weight, means that it takes up a lot of space. In automobiles it has to be compressed and stored in high-pressure tanks. The "fuel" tank would take up most of the space in the car. Compressing the gas takes a lot of energy in itself—an additional cost. To make a hydrogen fuel cell car with the same range as today's gasoline-powered car, with comparable passenger room, would require storing the hydrogen at 10,000 pounds per square inch (psi), which is ultrahigh pressure. This can be done, using ultrastrong carbon fibers to reinforce the tanks. Such a tank might even survive a high-speed crash. The question is whether the more delicate plumbing connections from the tank would survive. If not, hydrogen under extremely high pressure would escape rapidly. Hydrogen is extremely flammable. Hydrogen-and-air mixtures combust over a wide range of concentrations from 4 percent to 75 percent, and will be detonated by very low energy inputs less than one-tenth the energy needed to ignite air and gasoline. Because hydrogen produces considerable heat at decompression, it could self-ignite in a crash as the gas rushed out of a tank through broken valves.

Hydrogen presents two other problems for storage tanks. One is that it diffuses easily. That is, it leaks. Due to its extremely low atomic weight, it can escape through very small openings. It is very hard to contain. Hydrogen is also extremely corrosive. It likes to combine with other elements and compounds. The interior of the storage tanks, the pipe connections, the valves, and the seals would all be subject to much more rapid disintegration than is the case with gases such as methane. Also unlike gasoline, which is a liquid at regular air temperature, compressed gases are difficult to transfer from one vessel to another. To get hydrogen from stationary supply tanks to vehicle tanks would take additional energy.

Another category of concerns involves the transport of hydrogen to an infrastructure of "fuel" stations comparable to what America has developed in support of its current motoring system. Gasoline is distributed to stations in unpressurized tanks by trucks. Liquid hydrogen would have to be trucked around in ultra-high-pressure tanks. A loaded forty-ton tanker truck is designed to carry about twenty-five tons of gasoline. Because hydrogen is so light, a comparably sized tanker truck transporting hydrogen would only be able to carry about half a ton of hydrogen. The relative energy consumption by the truck compared to the energy value of its cargo would make hydrogen uneconomical at almost any distance.

Bossel and Eliasson write:

> A mid-size filling station on any frequented freeway easily sells 25 tons of fuel each day. This fuel can be delivered by one 40-ton gasoline truck. But it would need 21 hydrogen trucks to deliver the same amount of energy to the station, i.e., to provide fuel for the same number of cars per day. Efficient fuel cell vehicles would change this number somewhat but not considerably. The transfer of pressurized hydrogen from the truck to the filling station takes much more time than draining gasoline from the tanker into an underground storage tank. The filling station may have to close operations during some hours of the day for safety reasons. Today about one in 100 trucks is a gasoline or diesel tanker. For hydrogen distributed by road one may see 120 trucks, 21 [of which] or 17 percent of them transport hydrogen. One out of six accidents involving trucks would involve a hydrogen truck. This scenario is unacceptable for political and social reasons.[4]

Pipelines for distributing hydrogen around the United States would present more problems. The existing system built for natural gas just can't be used. It is composed of pipes that are not large enough, given hydrogen's extremely low density. Hydrogen would corrode the seals and destroy lubrication in the pumps that are necessary to keep gas moving through pipes at regular intervals over hundreds of miles. Its propensity for diffusion would result in unacceptable rates of leakage. In short, the existing pipeline network would have to be completely rebuilt in parallel for hydrogen

4. Ulf Bossel and Baldur Eliasson, "Energy and the Hydrogen Economy," EVWorld (*http://evworld.com*), January 2003.

at the cost of scores of billions of dollars (assuming the other technical problems could be overcome). This is unlikely to happen. Add to this that the infrastructure of every individual fueling station in America would have to be retrofitted.

What all this boils down to is that a hydrogen-powered automobile system and all its supporting infrastructure cannot be substituted for an oil-based system under any plausible equation currently understood. Without ubiquitous fuel stations and methods for supplying them that are both economically and logistically rational, there is no basis for such a system. This predicament underscores the very special nature of oil and the specialness of the systems that we have designed to run on it. It has powerful social implications as well. For instance, unless the hydrogen and fuel-cell system of personal transport was as democratically affordable to the broad public as the oil-based system was, how can we expect it to be politically acceptable? It has certainly been demonstrated that a fuel-cell car can be built—at least an expensive prototype. But what if it can be mass-produced only at a price that puts it in the luxury category for ordinary people? What if such cars can't be marketed for less than $80,000 (in 2005 dollars)? It suggests that a substantial portion of the public will not be able to participate in motoring. This would pose problems for a society in which cars have been virtually mandatory for the normal activities of daily life.

The longer you look at the particulars of a proposed "hydrogen economy," the more laughable a fantasy it appears to be. But it is instructive in showing the limits of our thinking, for instance, our blindness to other solutions for America's extreme car dependency in the coming permanent oil crisis. Instead of finding a new fuel to run suburbia, a far more sane and intelligent response might be for Americans to live in traditional walkable communities served by public transit. However, the psychology of previous investment, aggravated by our national mythology about individualism and country living, has so far prevented mainstream America from even considering this alternative. We've poured so much money into suburbia and its accessories that we cannot now allow ourselves to imagine giving it up. And the paradoxical bundle of ideas that combines the liberating nature of endless motoring with entitlement to a settled home in the rural landscape (the American Dream) still exerts awful pressures on our capacity to dream of other living arrangements. Americans who travel to Europe regularly and enjoy the life of the walkable city there also regularly vote against higher-density building proposals back home in Minneapolis and Nashville.

The upshot of all this is that there is not going to be a "hydrogen economy." We may use hydrogen in some new ways, and may continue producing commercial hydrogen chemical products. An enlarged nuclear power infrastructure may lower the cost of making hydrogen by electrolysis. But we are not going to run places like Hackensack, New Jersey, or Anaheim, California, on hydrogen. We are not going to replace the current U.S. automobile and truck fleet with hydrogen-powered cars. And, in the event that yet more miraculous technological breakthroughs occur that would alter the known laws of thermodynamics to make hydrogen as cheap as Texas oil once was, then there is going to be a Long Emergency between now and that rosy future.

Coal

Coal was the fuel that kicked off the industrial revolution. In England it was first grubbed out of surface pits and found along the seashore in places where waves had washed seams out of cliffs. It was hard to collect the stuff in large quantities and easier to chop down trees, if you owned some. Coal tended to be used by the landless poor who couldn't afford wood. Coal was regarded as inferior to wood for heating and cooking because of the smoke and odors it produced. Stoves and fireplace inserts that would allow it to be used comfortably hadn't been invented yet. As England's wood supply headed toward serious depletion in the eighteenth century, however, both rich and poor had less choice in the matter and turned increasingly to coal. The superior energy-producing characteristics of coal compared to wood mattered only after wood became relatively scarce and the equipment for burning it improved.

As coal became a more indispensable commodity, it became worthwhile to dig it more deeply out of the ground and trade it commercially. The coal pits became mines. The mines frequently flooded. Before long, the need to pump water out of coal mines provoked the development of coal-fired steam-powered pumps, which swiftly led to steam engines that could power boats, railroad locomotives, and manufacturing machinery—and England was off (as was America soon after). Coal was dirty and highly polluting, but it got so much work done that the pollution came to be tolerated as the cost of civilizational amenity. Despite killing "fogs" composed largely of coal smoke, there was no serious popular movement in London to stop the use of coal. By the twentieth century, though, wherever oil was accessible coal came to be phased out. Oil was easier to get out of the ground—especially in the early days—and much more versatile than coal. America led the way in the oil chapter of the industrial story, as the United States possessed plenty on its own territory, in places easy to get to, and developed that industry at the gigantic scale before anyone else.

Now as oil recedes into depletion in the twenty-first century, coal will very likely make a comeback. Most of the coal used in the United States today is burned in power plants to produce electricity. Coal currently produces about a quarter of America's electricity. By 2004, we were consuming about 1 billion tons a year. What coal does best is run stationary turbines, which are used in electric generating plants. Historically, coal was the first fuel of modern central home heating in America, and might have to be used that way again—though for a nation accustomed to clean, no-hassle, virtually automatic gas furnaces, conversion back to coal is apt to mean a marked loss of luxury. Coal in its normal solid form is obviously unsuited for that other major U.S. energy sink: the car.

We could run railroad locomotives with coal-powered steam engines, and perhaps we will have to resort to that, but it would make more sense to generate electricity to run the trains, if only because the particulate pollution and solid waste could be dealt with in fewer places.

Most of the happy talk about coal these days emanates from the coal mining industry. The industry says there is lots of coal left, enough to last hundreds of years. This remains to be seen. We have already mined much of the best-quality coal that was closest to the surface and easiest to get. A lot of what remains may be so hard to get that it won't be worth the energy expenditure necessary to get it. There is, in fact, a wide disparity of opinion about how much coal we will really be able to use. I don't doubt that we will have to resort to using coal to some extent when our problems with oil and gas depletion really kick in, but it isn't going to be cheap, the quality might not be so good, it isn't going to last that long, and it's not going to work nearly as well as gas and oil did. It depends partly on what we decide to do about nuclear energy, which I shall get to presently. If objections to nuclear power can't be overcome, then coal will be the logical candidate for generating the bulk of our electricity, at least for a while, if we want to keep the lights on.

Burning coal is still the greatest source of overall toxic air pollution in the country and probably a significant contributor to global warming. Coal burning produces a great deal of solid waste (5 percent to 20 percent) of its original volume. A single coal-fired electric plant can produce more than 1 million tons of solid waste a year. Coal produces 60 percent of the particulate emissions (cars and trucks produce most of the rest). Coal is implicated in mercury pollution that causes 60,000 cases of brain damage in newborn children every year in the United States. Coal is linked to asthma. Coal-fired power plants are chiefly responsible for acid rain. Perhaps these are prices that Americans will be willing to pay in order to enjoy high levels of electricity consumption in the future. It is certainly possible to clean up the emissions from coal-fired power plants, but it makes electricity more expensive, and the political will to clean up the industry may not be there in a more austere economy. In any case, even if heavy metals and particulates are taken out of the emissions, coal will still produce large quantities of carbon dioxide, the chief suspect in global warming. In 2003, the Bush administration, in fact, eased pollution standards for the power industry.[5]

Coal mining is also very destructive to the landscape and habitats. Surface and strip mining, now the most common method, levels whole regional topographies and poisons groundwater at prodigious rates. Resorting to coal as a major energy source would be a big step backward in the narrative of human progress. That doesn't mean it won't happen. The Dark Ages were also a step backward after the achievements of classical Rome, but they happened nonetheless. What we face may be more like the Dim Ages.

5. A new Environmental Protection Agency rule issued in August 2003 is a revision of the 1977 "new source review" provision, allowing the nation's most polluting power plants to upgrade equipment without implementing new emission-control measures. Scientists and officials have criticized it as the biggest rollback in the history of the Clean Air Act.

Hydroelectric Power

Hydroelectric power means electricity generated by water power, usually involving sites on rivers where moving water can be directed to spin turbines that activate a generator to produce electricity, or in reservoirs behind dams, where a reserve of water makes a constant, regular flow available from a river with an erratic or seasonal flow. Hydroelectric power can also be produced by harnessing tidal action, though that is a more difficult and expensive kind of project and done only on the grand scale.

Hydroelectric power is great. It's one of the oldest, most tried and true methods for making electricity. It doesn't produce carbon dioxide pollution (though the manufacture of hydro components does). The latest generation of turbines can operate at up to 90 percent energy efficiency. We've been using hydroelectric power in the United States since the first generating station on the Fox River in Appleton, Wisconsin, opened in 1882. Ten percent of electricity in the United States today comes from hydropower, compared to 40 percent in 1940. Hydro is well understood and fairly dependable. It can be done at scales from a one-household microgenerator on a creek to Hoover Dam, which lights up whole cities. Hydroelectric power is produced at about 2,200 sites recognized under the Federal Energy Regulatory Commission. All the great sites on major rivers in the United States have been exploited. They are problematic because soil and other material washing down a river deposits silt behind a dam, eventually rendering the dam inoperable. Virtually all the major dams in the United States are less than one hundred years old. All of them have sedimentation problems. Large reservoirs in the United States lose storage capacity at an average rate of around 0.2 percent per year, with regional variations ranging from 0.5 percent per year in the Pacific states to just 0.1 percent in reservoirs in the Northeast. Many of these still have a useful life of another century or more. But we will not be adding any major new dams and the total generating capacity will not grow in large increments. The U.S. Department of Energy (DOE) has identified 5,677 sites in the United States with undeveloped capacity of about 30,000 megawatts (MW). By comparison, today there are about 80,000 MW of hydroelectric generating plants in the United States. Virtually all these sites are on small rivers and creeks. That is actually good news, because future hydro projects will necessarily have to be on a smaller scale in a nation with a declining base of oil and gas resources and, more critically, far less government investment money. Small projects also imply that they will be geared to serving the localities where they are sited. This will be a good thing in a society far more locally oriented by circumstance.

Not all regions of the United States are equally endowed with running water. Those that are will be fortunate. My own area of eastern upstate New York, for example, a rugged terrain etched by small fast-flowing streams emptying into the Hudson River, is littered with decommissioned small hydroelectric stations. These were built aggressively in the first half of the twentieth century by local independent power producers to light up local towns, and allowed to go out of service after World War II as the bigger utility companies consolidated into giants. The giant companies, such as Niagara Mohawk, didn't want to bother maintaining the little stations. They were shut down and the equipment was sold for scrap. Some of the empty turbine buildings have been converted into houses in recent years. In the Long Emergency they may have to be converted back to hydroelectric stations.

If the DOE estimate of potential sites is correct, the United States could boost hydroelectric capacity by roughly 50 percent of its current level. Since hydro makes up only 10 percent of total U.S. electric generation, we would gain the equivalent of about 5 percent of total current usage if all the potential hydro sites were put to use. The total includes sites that might be considered environmentally sensitive and so a percentage of them might never be exploited. Hydroelectricity is good but utilizing it to the maximum will only fractionally compensate us for losses in natural gas generation that are sure to come. Hydroelectricity also raises a fundamental question that I will address below in detail: Can we even build the plants and equipment without an underlying base of cheap fossil fuel?

Our national system of interdependent giant regional distribution grids is widely considered to be in dangerously decrepit condition. This was underscored by the great regional blackout of 2003, which took out power from New York all the way to Detroit. The giant energy companies themselves seem to anticipate a major systems change to what they call "distributed generation," meaning that people will get their power closer to home. The trouble is, the big companies are far from confident about how that would be accomplished. There was a lot of excitement in the 1990s about the development of home fuel-cell generators. These units, each about the size of a refrigerator, would generate all household current as needed by means of fuel cells. Power lines could be dispensed with. One weak spot in the theory was that the fuel cells would run on natural gas, a commodity now in depletion. Another weak spot was that research and development by several companies, led by General Electric, had so far failed to engineer an affordable home generation unit. So distributed generation has come to naught so far. The upshot has been that the giant regional grids, with their long ranks of towers and power lines and substations, are not being maintained because the utility companies are still betting that they will be obsolete sooner rather than later. It may not be long, though, before a critical point is reached where the equipment will not be reparable. And in the Long Emergency we will certainly not have the financial resources to replace it. After that, all electric power may be local, and some localities will be luckier than others.

Solar and Wind Power

By solar energy we generally mean either *passive* solar construction techniques that allow buildings to capture sunlight in the form of heat and light, or the *active* conversion of solar radiation into usable electricity by photovoltaic cells. In the deeper sense, "solar" could also apply to fossil fuels, as they represent eons of solar energy stored in hydrocarbon compounds, and to everyday fuels such as cordwood and cow manure, which owe their existence to sunlight. But for the purposes of this discussion I am talking about the first two.

Passive solar power is great. You build something right and it keeps on giving back value in terms of comfort. Premodernist architecture was developed to take advantage of sunlight for heating and lighting buildings (and breezes for cooling, which are also produced by solar action on air). The development of these traditional techniques was a slow and painful accretion of experience over scores of centuries. It has only been the anomalous abundance of cheap oil and gas in our time that permitted builders, and especially architects preoccupied with style issues, to depart from traditional practices that took advantage of passive solar energy. The twentieth century was the era of glass "curtain walls" on office buildings, windows that didn't open (or didn't exist), titanium facades on civic structures, and other fashionable stunts for "dressing up" buildings to proclaim the bold creative genius of their designers. This kind of indulgent, narcissistic behavior was possible only in a cheap-energy society in which little mattered in architecture besides fashion and the status associated with being on the "cutting edge" of fashion. It didn't matter whether air or light got into a Frank Gehry-designed museum because that's what the air conditioning and halogen track lights were for. What mattered was that the city was blessed with a fashionable object created by a celebrity shaman. Alas, nothing is more subject to losing value by going out of date than something that is valued solely for being up-to-date.

Where houses were concerned the process was a little different, if only because the masses of the public overtly hated avant garde architecture, vastly preferring traditional-looking houses. The catch was that they were only traditional-looking in terms of cartoon ornament and massing. In all other respects they were actually very experimental, especially in terms of building materials and orientation to the natural elements. Materials such as styrofoam panelized cladding (brand name: Dryvit) created all kinds of problems with condensation and decay. House builders paid no attention to regional differences. The exact same model could be built in San Diego or Rochester, New York, without taking into account climatic variation, because cheap electricity split the difference. The rampant ugliness of the built landscape in America was high entropy made visible. Postwar houses in the southeastern states especially were able to dispense with all the traditional architectural appurtenances for managing uncomfortable weather—porches, high ceilings, transom windows—and the result was a species of phenomenally ugly air-conditioned bunkers utterly sealed off from the surrounding habitats.

You don't have to go to extremes to gain value from passive solar design. I once built a small post-and-beam house designed to soak up sunlight during the day and store it in a concrete slab. It was not a robust engineering effort in terms of energy efficiency. Yet I was able to keep the whole building comfortably warm on a winter day by firing up a small woodstove in the morning. It wasn't necessary to refire the stove until evening time. The heating bill was remarkably low. Running the house required very little work—seven minutes a day to cut kindling and another five to light fires in the stove. You might even figure in the one afternoon a year I had to spend stacking firewood delivered in a heap by dump truck. The house didn't even look weird, as more hyperengineered passive solar houses of that era did. In contrast, the stock products of the home-building industry in recent years have been ludicrous in terms of even minimally utilizing passive solar energy. The typical "McMansion," or super-sized tract house on a half-acre lot, with its "lawyer-foyer" and great room, is an energy hog and many of them may be uninhabitable in the coming age of energy austerity. They were designed under the assumption that natural gas would be cheap and plentiful forever.

In fact, the single-family stand-alone house may have a tragic destiny in the years ahead. For several generations this way of living has been the norm in America, but it hasn't always been so. The single-family house in a suburban subdivision owes everything to cheap energy and to the broad middle classes that cheap energy has made possible. Until the twentieth century, stand-alone houses in the rural setting were either farmhouses, villas, or peasant hovels. People who lived in a rural setting practiced rural lifeways, generally having to do with food production. People involved in trade, services, and labor lived in town, and proportionately far fewer of them were homeowners. I believe we will be heading back to that prior state. The twentieth-century single-family suburban home alienated from the surrounding landscape may soon be obsolete. The norms for housing in the coming era of energy austerity will have to be much more traditional and integral with their surroundings. Because we will have to grow more of our food close to home, land will be valued more for agriculture than for commuter houses. This profound shift in values will reestablish the distinction between country living and town living, with appropriate building typologies, and they will certainly require a return to passive solar building techniques.

Active solar power—solar electric generation—is another issue. Proven technology exists. It works, though not nearly as well as fossil fuel modes of power generation. I'm not sure that solar electric power can continue to exist outside the friendly confines of a fossil fuel economy. We know how to make photovoltaic cell arrays out of silicon, plastic, and metal, and we know how to manufacture storage batteries out of plastic and lead, and we know how to build charge controllers, inverters, and other devices for regulating the storage and flow of electricity—but can we make these things in the future without oil, gas, or coal? Maybe not. It takes a lot of energy, many barrels of oil, to manufacture deep-cycle batteries and solar panels, and it takes a platform of advanced systems—everything from metallurgy to plastics manufacturing—to mass-produce all the components and standardize their performance. I'm not convinced that active solar power may be anything but an interim stopgap in the Long Emergency that will follow the end of the fossil fuel age.

I have run a modest solar electric operation for four years at a remote Adirondack vacation house. We're off the grid there, unable to hook into any public utility power lines. We have four 50-watt solar panels feeding a six-cell deep-cycle battery bank connected to a 2,400-watt inverter, which turns the direct current (DC) from the batteries into alternating current (AC) that normal appliances run on. The system was chiefly designed to run a ½-horsepower AC electric water pump that lifts water from the lake to a pressure tank. The pump runs for a total of two or three minutes a day. We just don't use that much water. Otherwise, the system powers a laptop computer, a small stereo system, and half a dozen compact fluorescent light bulbs (which are never all switched on at a given time, but are used for tasks like dishwashing or reading). We do not run a refrigerator, because they use too much electricity. For a while we ran a propane-powered refrigerator but it was an old one and inefficient. So now we just bring bags of manufactured ice over from the mainland.

The electric system is nifty but very delicate. The batteries must be babied. I have to check them once a month with a manual squeeze-bulb hydrometer, to make sure that they are charging properly. It is a messy job and a little hazardous because the liquid inside the battery is sulfuric acid. I have to wear goggles to protect against squirts and splashes. During this operation, I have to add distilled water to each cell if needed.

The system worked pretty well for the first two years. We used whatever electricity we wanted within a certain boundary of paranoia about pushing it too far to the limit. We had plenty of water and enjoyed hot showers—courtesy of a propane-powered tankless water heater—and played rock and roll and had electric lights after dark. As it happened, though, in the summer of 2003, we had a very abnormal stretch of about six weeks without a single full sunny day—and many flat-out rainy days. Our site is far from ideal in any case, on a west-facing hillside, and even on perfect days direct sunlight hits the panels only after 12:30 in the afternoon. So, one morning in mid-July, after a solid week of rain, there wasn't enough juice left in the batteries to run the water pump. Our system stayed down for a good week. I just turned off the inverter, which itself uses electricity just being on, and waited for the sun to come out for a few days in a row. In the meantime we made do without running water, electric lights, rock and roll, or the computer.

The system cost about $3,000 in 2001. If we were hooked up to the grid, we wouldn't have used that amount of electricity at current prices in thirty summers—effectively for the rest of my lifetime. We didn't get it to save money. We got it because it was our only option for having some electricity at our summer place. As I said, it is a very modest system. If you were to run something closer to a normal American household on solar power—meaning a refrigerator, a clothes dryer (another energy-sucking devil), televisions, desktop computers, and so on—you would need something more like a twenty-four-cell battery bank running off sixteen solar panels. The hardware alone would run close to $20,000 (not including installation). The time needed to monitor and service the batteries would necessarily be greater, and batteries do go bad. Even with careful maintenance, the whole battery bank might have to be changed every ten years at the cost of thousands of dollars. The solar panels themselves would last quite a bit longer than the batteries, but even they are subject to ultraviolet degradation and exposure to water and ice. Of course, in certain regions of the country limited seasonal sunlight might make solar electric marginally worthwhile even if there were no alternative.

It is possible that improved batteries and more efficient solar cells may be engineered. So far, however, the battery problem has been particularly vexing. The technology has not changed much in nearly a century. The lead-acid wet-cell batteries in my circa-2001 solar electric system are not substantially different from the battery in a 1912 Oldsmobile, and although researchers have been working doggedly in recent years to improve battery technology, their work has yielded only modest refinements. For example, lithium-based batteries work well in laptop computers and

LED lights, but so far they have not been economically scalable for household solar power systems. This is one of the main reasons that electric cars have been such a flop during the past decade: The batteries could not be improved to make them significantly less bulky or lighter, or to increase the travel range between charges. What's more, electric cars would have carried a base price 30 percent higher than comparable gasoline models, while the batteries would have to be replaced every few years for many thousands of dollars more. These problems left the electric car in oblivion. But they were developed in the first place not in expectation of oil shortages but to mitigate the separate problem of air pollution. In 2001, the California legislature mandated that 10 percent of all cars sold in the state be low-emission vehicles by 2003. In 2003, having failed abysmally to interest the public in buying electric cars, California rescinded the mandate. Meanwhile, General Motors shelved the development of its once-touted EV (electric vehicle). As of late 2003, both Ford and General Motors were turning their attention to fuel-cell cars instead—the idea being that a fuel-cell car would be in effect an electric car, using an electric motor, only without the bothersome batteries. However, fuel-cell cars are problematic for reasons already discussed pertaining to hydrogen and natural gas.

There is a set of erroneous popular notions to the effect that renewable energy systems such as solar power, wind power, and the like are available as freestanding replacements for our fossil-fuel-based system, that they are pollution-free and problem free—that renewables represent something akin to perpetual motion, a gift from the sun. The operation of a solar electric system, like the one I run on an Adirondack lake, does not itself produce pollution, but the manufacturing of the components certainly does. The batteries, the panels, the electronics, the wires, and the plastics all require mining operations and factories using fossil fuels. And the components were transported by diesel truck to the marina dock from far away, and ferried to the site via motorboat. This gets back to the question as to whether these systems could exist without the platform of an oil or coal economy to produce them.

I don't think so. And in the absence of fossil fuels, what else? It is not at all clear, for instance, whether nuclear energy could be employed to manufacture solar components, as nonmilitary nuclear power has been used solely to generate electricity, not large-scale industrial procedures.

Might it be possible? Nuclear fission can produce plenty of heat. That is one of the reasons that reactors can be so dangerous. But there's no precedent for nuclear reactors being used in direct manufacturing processes, except to make other radioactive materials.

Solar electric and wind power therefore might be viewed as accessories of the fossil fuel economy.

The arguments for and against wind power are very similar. Wind power presents some possibilities that solar power does not. The energy captured by wind turbines can be captured or stored in ways other than electric batteries, especially during those times when a wind "farm" (a collection of windmills) is producing a surplus beyond what customers are using. One possibility is pumping water up into storage reservoirs to operate hydroturbines in offline periods. But this depends on favorable topography. It wouldn't work in Nebraska. And a substantial amount of energy would be lost in the conversion process. An idea along similar lines would involve the injection of compressed air, or other gases, into salt caverns or aquifers from which the energy potential can be recaptured to drive generating machinery. Good sites for underground compressed air storage are an issue. And to achieve efficiency, the compressed air has to be used in a partnership with natural gas. Compressed air/natural gas turbines are three times as efficient as conventional gas turbines—but the system implies a reliable supply of affordable natural gas and the United States is past peak and depleting fast. It is conceivable that wind power could be used to produce synthetic methane by re-forming carbon dioxide in the presence of a catalyst under heat and pressure. But like other alternative fuel schemes, it raises issues of economy and scalability. Can the infrastructure of the United States, as presently configured, be run on these things? No way; not even a tiny fraction of it.

The wind power inquiry eventually would lead back to the same place as the one on solar power: Can these technologies be detached from the fossil fuel platform supporting them? Sure, it is possible to generate electricity using wind turbines. Yes, European nations have made major investments in "wind farms." Denmark was getting 18 percent of its electricity from wind in 2003, the most per capita of any country. Germany was producing more than 10,000 megawatts from its installations, Spain more than 3,000. This is all possible because the world has been at or around the historic peak of oil production, meaning the oil economy at the millennium was at its most robust just when these wind farms were set up. Thanks to fossil fuels, you could produce the special alloy metals needed to make the turbines, and you could run factories to mass-produce them and make the replacement parts—because wind turbines are notoriously finicky and break down a lot—and you could set up the installations using petroleum-powered heavy equipment, backhoes and front-end loaders and bucket trucks and what-have-you to prepare the site and jockey the machines into place. What happens without the fantastic technological support of the oil economy in the background?

The advanced industrial nations need to have all the necessary alternative energy infrastructure in place long before that background support disappears. Still, it begs the question about anything beyond the short-term future. The advanced nations could consciously commit themselves to dedicating some portion of the world's remaining oil endowment to the production of wind turbines, solar arrays, and batteries—but don't count on it happening. American leaders haven't paid attention to energy issues since the oil crises of the 1970s. It's hard to believe that we are suddenly going to behave more intelligently. Anyway, most of that remaining oil is not under American control. We are already fighting over it.

What happens when the people of the world are locked in conflict over the remaining oil? That is going to strain the relative international order that has allowed the global economy to function smoothly—order that we have taken for granted. It could give way to an international climate of military strife, mutual suspicion, and other discontents that would scuttle the global cooperation in finance and trade that we have come to depend on. Supply lines might be suspended or interrupted. How do we get exotic ores, chromium, titanium, from the few places that possess them to the foundries where the alloys are manufactured in order to manufacture wind turbines? What do we use to power the furnaces? Coal? Coal is generally mined using diesel-powered equipment. Well, artificial diesel fuel can be made from coal, or one can reinvent coal-powered steam shovels and the like, but it would be necessary to ramp up whole new industries on a shrinking petroleum energy base. Then what happens when the coal runs out? The coal industry predicts that the U.S. coal supply will last for a couple of hundred years. Historically, that's rather brief, comparable to the time between Cortes's conquest of the Aztecs and the birth of Ben Franklin. That's in a best-case scenario. More likely, the tail end of the coal supply will be the coal that is the most difficult to get out of the ground, in the worst places, with the lowest return in energy invested (ERoEI), and possibly not even retrievable at the current scale of mining.

My larger point is that the high-tech gadget-oriented vision of "renewable" energy, as imagined by the most wishful, rests on the quicksand of diminishing returns. There would seem to be a parallel belief among a pragmatic subset of the wishful that the tactic of using the remaining fossil fuels to prepare for a post-fossil fuel future is a matter of buying time until "they," the scientist-nerd-innovator-geniuses, *come up with* a new and superior energy source. For all I know, this miracle will occur. Weirder things have happened in human history. (What would Ben Franklin have thought of Adobe Photoshop?) However, this idea of buying time until the tech demigods deliver a technology miracle is just another way of describing a cargo cult. From the standpoint of group psychology, it puts the human race into a jam, cramming for a final exam that it can't afford to fail. And, as if this were not bad enough, other forces and circumstances that I'll discuss presently, such as climate change and the spread of disease, also beg the question as to just how bad this jam is—whether we've exceeded (and actually violated) the carrying capacity of the planet so egregiously that no ventures into alternative energy will allow us to keep the current game going.

While the demise of fossil fuels may deprive us of some kinds of technology we have become used to, it may or may not entail a loss of technological knowledge. The Romans developed the technology of building in reinforced concrete to an extremely high level of refinement—and an artistry in working with it to match. That knowledge was lost for more than a thousand years after the fall of the empire. The great cathedrals of medieval Europe, for all their majesty, represent a much more primitive technology—the mere gluing together of stones with mortar—than the construction of something like the Pantheon a thousand years earlier, which employed progressively thinner courses and lighter blends of concrete from the base to the top of its dome. That level of technology was not recovered until the early twentieth century, and the process of first acquiring, then losing, and then regaining the knowledge had as much to do with social and economic organization as it did with the possession of sheer technological information. Roman architecture would have been impossible without the complex socioeconomic platform of empire. The medieval social platform for northern European life was less elaborate and arguably less complex. Compare these two historical cases with the complexity of social and economic organization that allows oil to be extracted from the ground, refined to gasoline, transported six thousand miles, and used in a highly engineered, fine-tuned machine called a car, driven on a six-lane freeway. If the social and economic platform fails, how long before the knowledge base dissolves? Two hundred years from now, will anyone know how to build or even repair a 1962 Chrysler slant-six engine? Not to mention a Nordex 1500 kW wind turbine?

We currently possess enough knowledge to employ and optimize the lower-entropy activities of the future, or at least recognize the futility of attempting to sustain the unsustainable in our current high-entropy mode of life. The existing knowledge in basic physics and chemistry is so widespread that it is likely to persist quite a while into the future and provide a foundation for doing more with less than, say, the people of the eighteenth century were able to do with their more limited knowledge. I'm not proposing that we simply go back to a preindustrial mode of living. Modernity itself has already led to tremendous loss of knowledge of sustainable living practices that were followed for thousands of years.

There are other ways of using the sun and the wind that don't rely on high-tech gadgets such as solar arrays and turbines, and we will rely on them much more in the years ahead. A draft horse is a solar-powered farming implement that is capable of reproducing, i.e., a self-renewable. It implies, however, an entirely different system of agriculture. Home gardening is a solar-powered activity that produces food at the family scale. In our time, home gardening has degenerated to little more than exterior decoration. We'll surely have to grow more of our food closer to home in the Long Emergency, and those of us who have any land at all, even a yard in a city house, will do so. Wind, solar, and water power can do a lot of useful work at the small and medium scale, unmediated by fossil fuels. We will certainly have to do more with them at the small, local scale in any kind of plausible future.

Fossil fuels allowed the human race to operate highly complex systems at gigantic scales. Renewable energy sources are not compatible with those systems and scales. Renewables will not be able to take the place of oil and gas in running those systems. The systems themselves will have to go. Even many "environmentalists" and "greens" of our day seem to think that all we have to do is switch inputs. Instead of running all the air conditioners of Houston on oil- or gas-generated electricity, we'll use wind farms, or massive solar arrays; we'll have super-fuel-efficient cars and keep on commuting over the interstate highway system. It isn't going to happen. The wish to keep running the same giant systems at gigantic scale using renewables is the heart of our illusions about solar, wind, and water power.

Synthetic Oil

Coal can be processed into very high-grade synthetic oil and gasoline, as it is itself just a solid hydrocarbon version of the same prehistoric organic goop from which oil was formed. The Nazis were able to do a lot with coal during World War II. They had to because they possessed almost no oil of their own. But they had rich supplies of coal. In the 1930s, when the United States was getting half its total energy from coal, Germany was still getting 90 percent from coal—and only 5 percent from oil. When Adolf Hitler came to power in 1933, he had already enlisted the help of the giant chemical company I. G. Farben in a scheme to produce significant quantities of synthetic oil from coal.[6] The process had been invented in Germany in 1913 by Nobel Prize-winning chemist Friedrich Bergius, and I. G. Farben owned the patents. It involved adding hydrogen to coal under high temperature and pressure, in the presence of a catalyst. The process was energy-intensive and expensive, but price was no object for Hitler. By September 1939, as he prepared to invade Poland, Germany was running fourteen hydrogenation plants for making synthetic gasoline and aviation fuel, with six more on the drawing board.

6. Yergin, pp. 328, 329.

Coal would supply about half the liquid fuel needed by Hitler's armies in the coming world war. The balance in conventional oil came, at first, from Romania and Russia. But Hitler did not like being fuel-dependent on the Bolsheviks he so despised. Eventually, he turned his sights on capturing the Soviet oil fields around Baku, and indeed that is why in 1941 he broke the 1939 antiaggression pact with Stalin and launched Operation Barbarossa, the invasion of Russia that would begin his undoing. The failure of the Russian campaign, or to secure the Romanian oil fields, left the Germans desperate for fuel to keep their war machine running. They managed, amazingly, to keep on producing enough synfuel, despite the massive allied bombing campaign against German industry, to nearly beat back the American advance in the Ardennes in December 1944. But the following spring, the Nazi war machine literally ran out of gas, and that was that.

Years later, with the war and Hitler and Nazism long behind us, the memory of synthetic fuels lingered on. President Nixon turned to "synfuels" in the wake of the 1973 OPEC oil embargo—at least the idea appealed to him because it could be neatly packaged for political consumption when he was otherwise floundering in the Watergate swamp. Of course, it was one thing for Nazis to wring gasoline from coal under wartime conditions in a nationalized economy using vast amounts of slave labor, and quite another to do it in a free country on an economically sound market basis. Despite the tremendous paranoia and economic havoc induced by the 1973 oil crisis, no coal synfuel plants were constructed in the wake of the OPEC embargo. Nixon's successor, Gerald Ford, proposed government support for a more specific program that would set up twenty plants to produce a total of 1 million barrels of synthetic fuel a day. (The United States currently consumes around 20 million barrels of oil a day.) Ford's bill did not make it through Congress. A few years later, in July 1979, President Carter proposed an $88 billion decade-long effort to promote production of synthetic fuels from coal and shale oil. Carter was haunted by the energy predicament. As a trained naval nuclear engineer, he could read the trends in America's energy future. He had entered office with the nation still shuddering from the aftereffects of the OPEC embargo and his anxieties were confirmed with a second oil crisis brought on by the fall of the shah of Iran. Unfortunately Carter was ahead of the public, who merely viewed all the machinations around oil as perfidy by Arabs or (interchangeably) greedy oil companies. Carter tried to persuade them that the predicament was for real, "the moral equivalent of war," but he was widely ridiculed for his efforts.

Carter's successor, Ronald Reagan, canceled the synfuels initiatives altogether because he believed that there wasn't an energy problem that couldn't be solved by deregulation and government leaving free enterprise alone. Reagan was lucky. In the middle of his two terms, the bottom fell out of the oil market and prices commenced a fifteen-year-long slump, brought on by a number of factors: by the full-throttle production of a fraying Soviet Union trying to desperately acquire hard currency and stave off collapse; by the fruits of ramped-up oil exploration commenced after the crises of the 1970s, including the North Sea bonanzas of Britain and Norway; and by the unraveling of OPEC's price discipline brought on by the overproduction of desperate member nations such as Nigeria and Venezuela. All these things put a lot more oil into the global market pool and drastically depressed the price per barrel for more than a decade, from 1986 until 2001.

The first President George Bush was therefore able to ignore the energy issue, except as it manifested in international affairs: the first Gulf War following Iraq's invasion of Kuwait, which was caused, in part, by Kuwaiti cheating—drilling horizontally beyond its borders into fields under Iraqi territory. In the meantime, Bush did nothing to revive the synthetic fuels program. His successor, Bill Clinton, served during the heart of the 1990s oil glut, as the North Sea fields pumped at full throttle and world production continued to increase just short of the historic peak, and a relative if fragile global peace prevailed. Prices for oil continued to sink toward a postwar low. Clinton, the archetypal yuppie suburbanite, did nothing to prepare the nation for the post-peak era and enjoyed the luxury of ignoring energy issues generally—while the nation outsourced its manufacturing capacity and a "new" economy based on suburban sprawl land development stealthily took its place. George W. Bush, the second President Bush, had the misfortune to be in the White House as the global peak event neared and the oil markets began to wobble. Bush and his vice-president, Dick Cheney, both former oil executives, addressed only one of the manifestations of it—Islamic fundamentalist terrorism—by engaging in the first phase of what was likely to be a long war to control and pacify the Middle East. By early 2005, Bush had done nothing meaningful about energy policy in general and synfuels in particular.

This record of inaction around synthetic fuel over the past thirty years by public or private means or any combination would seem to beg a fundamental question: Do synfuels make sense in anything but a wartime emergency? My guess is that they will not. Propagandists from the coal industry claim that the cost of oil made from coal has come down from about $50 a barrel in 1973 to $30 in 2003, but when the price of natural crude rose into the $50 range in the fall of 2004 there was no fanfare to announce any new ventures in synthetic oil by the coal industry. When the time comes that global oil allocation becomes permanently disrupted, coal producers are counting on the likelihood that Americans will be desperate enough to pay whatever is necessary for coal-derived liquid fuels.

The money has to come from somewhere, though, and if Americans are spending proportionally more money fueling cars and trucks, then other things that define our standard of living will suffer. I think we can state pretty categorically that an economy without reliable supplies of cheap oil is going to be much weaker, generating less business activity, creating ever more economic losers who are not going to be able to afford either the synthetic fuels or the cars to run them. In other words, the fact that it is possible to make oil out of coal doesn't mean that it can economically replace cheap and reliable supplies of natural oil to run the American Dream. Like hydrogen, synfuel can be manufactured, but it doesn't scale upwardly.

The only plausible application of coal-derived liquid synfuels will be in the military, and even that is debatable. If the current wars for control over the Middle East continue for a long time, as they are likely to, or if they spread to other oil-producing regions, or if they don't go well for our side, the United States may more and more find itself in a squeeze similar to Germany's six decades earlier. Before that happens, though, the civilian sector would be subject to brutal gas rationing that would make the American Dream suburban way of life very difficult to carry on and would undermine the ability of the nation to prosecute war for oil, or for freedom, or for anything else.

In the early years of the twenty-first century, what scant synfuels activity there was amounted to little more than a corporate tax credit scam. Coal that was "chemically modified" qualified for significant tax credits. The law didn't spell out what "chemically modified" meant exactly. So some sharp corporate lawyers and techno-nerds cooked up a process whereby coal could be sprayed with small amounts of diesel oil, pine tar resin, and other substances, and sympathetic IRS higher-ups ruled that the product was a synthetic fuel. It was that simple. Companies engaging in this legalized scam aren't even coal companies—for example, the Marriott hotel chain, which acquired four coal "synfuel plants" in October 2001. To call them "synfuel plants," though, is a comedic stretch. They consisted merely of a few pole barns and conveyer belts where coal was sprayed. The following year, 2002, the first full year of operation, Marriott generated $159 million in tax credits from spraying coal with oil and other materials. The company had paid only $46 million in cash for the facilities, meaning that the tax credits gave the company a return of 246 percent on its investment in just one year—at a time when room revenue from the hotel side of the business had fallen 4.8 percent. What's more, the company's effective income tax rate plunged to 6.8 percent in 2002 from 36.1 percent in 2001, "primarily due to the impact of our synthetic-fuel business," according to its annual report.

Thermal Depolymerization

There was a big stir in energy circles in the spring of 2003 when *Discover* magazine published a splashy article titled "Anything Into Oil."[7] A company called Changing World Technologies, with a plant in Missouri, claimed that it could take any carbon-based feedstock imaginable, "including turkey offal, tires, plastic bottles, old computers, municipal garbage, cornstalks, paper-pulp effluent, infectious medical waste, oil refinery residues, even biological weapons such as anthrax spores," and convert them into three valuable products: high-quality hydrocarbon oil, clean-burning hydrocarbon gas, and useful minerals. They called it thermal depolymerization, or TDP. It was a high-tech method for mimicking and greatly accelerating

7. Brad Lemley, "Anything Into Oil," *Discover*, Vol. 24, No. 5 (May 2003).

the process that nature followed in creating geological oil out of fossil organic remains. The article claimed: "If a 175-pound man fell into one end, he would come out the other end as 38 pounds of oil, seven pounds of gas, and seven pounds of minerals, as well as 123 pounds of sterilized water." Oil derived from, say, turkey guts, would chemically resemble No. 2 fuel oil of the kind used in home furnaces. Engineers and investment bankers alike joined in the cheerleading. The federal government kicked in $12 million in research grant money to the project.

The machinery used resembles that in a conventional oil refinery—on a much smaller scale. The company claimed the process was 85 percent energy efficient for inputs such as turkey guts, meaning that for every 100 BTUs produced out of feedstock, it took only 15 BTUs to run the process. Water in the wet slurries, like the turkey guts, was ingeniously enlisted to help in phase one of the process, a primary "cooking" at 500 degrees Fahrenheit and 600 psi pressure, converting fats, proteins, and carbohydrates into carboxylic acid. When the pressure was dropped rapidly, about 90 percent of the free water was driven off. This did away with having to remove the water by heating and evaporation. The second stage of the process broke the hydrocarbon chains down further, turning them eventually into a light oil. The third phase worked like a conventional oil distillery. Hydrocarbons were separated by molecular weight into kerosene, gasoline, naphtha, and so on. The captured flammable gas was used in turn to fuel the process.

Dry feedstocks such as PVC plastic from ground-up appliances and building materials would be mixed with water for processing to yield useful chemicals such as hydrochloric acid, as well as hydrocarbon fuels. The different feedstocks required different "recipes" and cooking times. Changing World Technologies claimed it could safely reprocess anything short of nuclear waste. The company's first commercially scaled plant, a $20 million installation in Carthage, Missouri, was built next to a ConAgra Foods Butterball Turkey processing factory. Company spokespersons claimed that they would ultimately make oil by this method for $10 a barrel in 2003 dollars.

Anything that sounds too good to be true usually is, and that is certainly the case with thermal depolymerization. It has overtones of those perpetual motion schemes from the nineteenth century. Garbage in, oil out. (And there will always be plenty of garbage, right?) In fact, it amounts to a recycling program. TDP takes items produced by our high-entropy abundant-oil economy and turns them back into oil with (supposedly) only a modest 15 percent energy loss as a nod to the second law of thermodynamics (the "law of entropy"). The catch is that the oil economy platform has to be there in the first place. For instance, those gigantic turkey breeding operations run by ConAgra are possible only in an agricultural system run on cheap oil and natural gas, in particular to manufacture fertilizer for raising the grains fed to the turkeys, but also for housing, processing, freezing, transporting, and marketing the birds at a gigantic scale of enterprise that ends in the freezer section of a 150,000-square-foot mega-supermarket. Without fossil fuels, turkey farming would have to take place on a much smaller scale on a much more localized basis, and the amount of waste products in the form of feathers, guts, and feces would not be worth firing up even a demonstration-sized TDP distillery. (And if you were to travel around collecting all the hypothetical turkey offal from many distributed local turkey farmers, and take it to a hypothetically centralized TDP plant, the gasoline or diesel fuel expended in the collection effort might cancel the oil gained from the turkey offal.) The second law of thermodynamics never rests.

A similar picture resolves concerning all the other putative "feedstocks" of the TDP process: tires, plastic bottles, old computers, municipal garbage, and so on. All of these things exist because they are products of abundant oil. Remove the cheap oil and sooner or later you have no feedstocks. TDP may be an excellent, efficient method for dealing with extant garbage and waste under current conditions. But current conditions have a short horizon. At some point fairly soon, the underlying platform of our fossil fuel economy is going to totter. When it does, we are not going to recover enough oil from the waste, junk, and garbage left lying around to keep running this way of life for any significant amount of time. Even if all the garbage produced every day in the United States under current conditions was converted to oil by TDP it would not amount to even 5 percent of our daily consumption of oil. One conclusion you could draw, then, is that if we reduced our energy use by 95 percent, TDP might work for us—but then if we reduced our energy consumption that much, we wouldn't be producing the amounts of garbage required to produce even that measly 5 percent.

Biomass

Forget biomass. It's only a cruder variation of thermal depolymerization. The idea is that we would supplement our fossil fuel-burning power plants by adding organic materials such as cornstalks, switchgrass, willow sticks, and sawdust. Biomass schemes are predicated entirely on the assumption of an underlying fossil fuel platform, especially in terms of agricultural waste products such as cornstalks grown under an industrial agriculture regime using massive petroleum and natural gas "inputs" for artificial manufactured fertilizers, harvesting, and transport. This applies in particular to all schemes promoting ethanol (alcohol derived from plants) as an "environmentally friendly" additive to gasoline. The amount of petroleum and natural gas needed to produce the corn to make the ethanol would more than cancel out any benefit from using a supposedly non-fossil fuel.

In fact, we will surely have to resort to one particular form of "biomass" use in the future, but not in any way resembling the fantasies proposed by the corporate and environmentalist tech-meisters. That is, we'll probably have to burn a lot of wood to stay warm in the Northern Hemisphere, which means that many of us in advanced industrial societies will be returning in some respects to preindustrial modes of living. In this event, I think we can expect a fairly massive devastation of forest in those places—such as America east of the Mississippi—where forests had been able to recover during the many decades when coal, oil, and natural gas reigned in home heating. The future deforestation of North America (and Europe) could be as rapid and dramatic as the extermination of the American bison in the decades after the Civil War.

Methane Hydrates

An immense amount of methane, natural gas, equal to at least twice the amount of all known fossil fuels on earth, is thought to be trapped in the ocean sediments as a gas hydrate. This is a kind of "ice" consisting of methane molecules, each surrounded by a "cage" of water molecules, stable only at low temperatures and extreme pressures typical of water depths below about a thousand feet. They do represent a possible recoverable energy resource, but with important reservations. One is that methane hydrates are very difficult to recover, meaning expensive, meaning they may require more energy to get than they produce when recovered, meaning they would be basically uneconomical. In fact, to date exactly zero methane hydrates have been commercially recovered.

Methane hydrate is also hazardous. Attempted underwater "mining" operations so far have led to explosions, including the destruction of drilling platforms and ships. The physical properties of methane hydrate are such that any attempt to recover them tends to wildly destabilize the material, causing the water and methane to dissociate. The freed highly flammable gas then rises to the surface. Industry has concerns about drilling through hydrate zones, which can destabilize supporting foundations for platforms. The disruption to the ocean floor also could result in surface slumping or faulting, which could endanger work crews and the environment. In addition to posing danger to humans attempting to mine it, methane freed into the atmosphere is a ten times more effective greenhouse gas than carbon dioxide. Released in any quantity, it would accelerate the problem of climate change. So far, attempts to recover methane hydrates have resulted in releases of methane into the atmosphere proportionately much greater than the gas recovered in the process.

Zero-Point Energy (ZPE)

This is an arcane process posed theoretically by quantum physicists. It has been called "the ultimate quantum free lunch." ZPE claims to be a theory for harnessing the energy potential of the "dark matter" of the universe. The dense and abstruse physics surrounding ZPE appears to contend that cosmic forces responsible for gravity can be accessed for unlimited supplies of cheap, pollution-free energy on earth. These theoretics are beyond the competence of the author, and so I will make only two points about ZPE: (1) A useful maxim in engineering states that when something sounds too good to be true, it generally is not true. This has been the case classically with perpetual motion devices and other claimed fantastic inventions such as internal combustion engines that can run on water and special carburetors that will allow an ordinary car to get two hundred miles per gallon. For now, ZPE seems to fall into that category. But who knows? One might have said the same thing about atomic energy in 1893. (2) If there is anything to ZPE, it is not likely to see practical development before the world finds itself in deep trouble over depleting hydrocarbon resources, if ever. One also must wonder, as in the case with other alternative energy systems, whether development of something like ZPE can occur absent an underlying fossil fuel technology platform to support the necessary work.

Nuclear Energy

Since the so-called "alternative" energy sources described above are all in one way or another implausible on a long-term basis without the subsidy of oil, the only remaining alternative is nuclear energy. About 20 percent of the electricity generated in the United States today comes from plants powered by nuclear reactors. In France it is closer to 70 percent (much of the rest is hydroelectric). Despite the fact that the use of nuclear power has become rather routine, it is extremely problematic in the long term for reasons that go far beyond, but include, plain energy economics, and it is fraught with potentially great political tribulation. But in the short-to-medium term, it might be all we really have to fall back on.

What the nuclear option comes down to is this: Unless we want living standards in the United States to slide far beyond premodern levels in the absence of cheap oil and natural gas, we will have to use nuclear fission as our principal method for generating electricity for some time into the twenty-first century while we scramble to make other arrangements. However, even if the United States embarks on an aggressive policy of building a new generation of nuclear reactors, life will still have to change drastically. It's really a question of whether we want these changes to happen with the lights on or the lights off. What distinguishes modern life most from premodern life is our access to electricity, and especially liberal, regular supplies of it.

We surely will have to reform our land-use habits and the oil-based transportation system that has allowed us to run our car-crazy suburban environments. We'll have to drastically change the way we grow our food and where we grow it. Social organization may be quite different in the decades ahead. Features of contemporary life that we have taken for granted, such as commercial aviation and canned entertainments, may fade into history. Politics that evolved to suit the fossil fuel fiesta, both on the right and the left, may morph beyond recognition around new forms, patterns, and values. But if we want the enterprise of civilization to continue as a general proposition, we'll have to keep the lights on, and the only way to do that by the mid-twenty-first century will be by using nuclear reactors to generate electricity.

I am not entirely convinced that we can do this for long without the fossil fuel platform to support the construction, manufacture, maintenance, mining, and processing activities that are necessary to create and service nuclear reactors. But the power obtainable from nuclear fission is so much greater than that of solar-electric, wind, biomass, and all other "alternative" fuels that an investment of any remaining fossil fuel in nuclear power could be more than a break-even or dead-loss proposition, and might, in turn, buy the human race more time to make more sustainable arrangements. Thirty years from now, we may have to resort to coal to service nuclear reactors, or perhaps synthetic oil derived from coal. But the basic energy equation of nuclear power vis-à-vis coal is very plain: One single atom of fissionable uranium will produce 10 million times as much energy as the burning of a single carbon atom. Uranium will produce 2 million times as much energy per unit mass as oil.

There is enough naturally occurring conventional uranium around to generate electricity based on current technology for perhaps a hundred years. Naturally occurring uranium is composed of two isotopes: It is 99.3 percent U-238 and 0.7 percent U-235. U-235 is the more fissionable of the two. Most nuclear power plants today use enriched uranium, in which the concentration of U-235 is increased from 0.7 percent to about 4 to 5 percent. Uranium is relatively cheap—about $30 per kilogram (2.2 pounds). The amount of uranium needed to supply electricity for a family of four for a lifetime would fit in a beer can.

There are 109 licensed nuclear power reactors in the United States and about 400 in the world. Reactors work by producing heat from controlled nuclear fission—that is, from neutrons induced by a critical mass of uranium atoms bombarding adjacent nuclei and splitting off more neutrons, which do the same. As the neutrons fly around, the content of the atoms changes and the original elements are transformed into other elements. The process generates enormous amounts of heat. The heat is used to create steam, which drives electric turbines. So, past the reactor core, the process is not much different than making electricity by any other steam-driven method. The process does not produce any of the gases associated with air pollution—no carbon dioxide, no ozone, and so forth. However, the activities necessary to construct and maintain a reactor certainly do produce plenty of polluting gases. The reactor waste itself contains hundreds of exotic, poisonous, radioactive toxins that were not found on earth before the advent of artificial nuclear fission.

The fuel rods in the most common reactors contain pellets of enriched uranium. The critical mass of fissionable material is adjusted by raising and lowering these rods inside the reactor core. Roughly every two years (to simplify it a bit), the fuel rods in a reactor become "spent" and have to be changed. It's a process that must be done carefully and can often take months, although improved methods have reduced the time, in some cases, to a few weeks. The spent fuel rods are still dangerously radioactive and just plain hot. The most vexing problem with running nuclear power plants has been the disposal of spent fuel. It has been more a political problem than a true logistical problem. Nobody wants such a waste storage facility anywhere near them. (Of course, nobody wants to live with their lights permanently off, either.)

Until recently, the designated national nuclear storage facility at Yucca Mountain, Nevada, a set of deep salt caves, could not be used because of the fear of offending constituent groups or environmental watchdogs. The site was below the old decommissioned U.S. atomic testing ground. A 1996 earthquake in the region had reinvigorated the fight against using Yucca Mountain, which was only a hundred miles from Las Vegas. There were fears in particular that radioactive material could invade deep groundwater and spread all over. Therefore, most of the spent fuel rods of American reactors have been stockpiled at reactor sites all around the nation, in storage vessels that resemble swimming pools, where the material becomes steadily less radioactive as the more unstable isotopes decay and also generate less and less heat. This method of stockpiling onsite has always been considered a temporary stopgap, but has become routine pending the resolution of a national nuclear waste storage program. Spent fuel rods can also be reprocessed in such a way that enough fissionable material is recovered from one batch to run a given reactor for an additional year. Ultimately, though, the waste has to go somewhere and has been accumulating all over the nation for decades. The average reactor will produce about 1.5 tons of waste per year. When incorporated in a stable glass matrix, this would amount to around five cubic yards of waste. Since the first commercial nuclear power plant began producing electricity in 1957, the total amount of accumulated spent fuel is 9,000 tons. It would all fit inside a space equivalent to a high school gymnasium with room to spare.

In July 2002, President George W. Bush signed House Joint Resolution 87, allowing the U.S. Department of Energy to take the next step in establishing a safe repository at Yucca Mountain. The DOE is currently preparing an application to obtain a Nuclear Regulatory Commission license to proceed with construction of the repository. This has ended the long political deadlock, though not the profound questions of ultimate safety. It takes five hundred years for the spent, stored waste of a nuclear reactor to decay to the point at which it is only as dangerous as naturally occurring uranium ore.

In reality, there may only be such a thing as *relative* safety. But it is worth considering that many more lives have been lost in the coal industry than in the nuclear power industry in the past five decades. In the past forty years, not a single fatality has occurred as a result of the operation of a civilian nuclear power plant in the United States, Western Europe, Japan, or South Korea. The Chernobyl nuclear power plant accident on April 26, 1986, in the former Soviet Union, was another matter. Thirty-one people died as a direct result of the explosion and fire that followed. The largest estimates of cancer deaths related to the Chernobyl accident is in the low thousands, with an unknown number of cancer cases yet to present in people who were children at the time of the explosion. About twenty square miles of land became uninhabitable for a long time. In comparison, there were no deaths in the 1979 accident at Three Mile Island, Pennsylvania. Radioactive gases were vented, but there is no accepted evidence that this harmed the public.

The Chernobyl reactor was a Russian-designed RMBK model infamous for its built-in lack of safety features. It was designed in the spirit of

Soviet expedience to both produce electricity and make bomb-grade material at the same time. The reactor didn't have a containment shell. It was also designed in such a way that if the reactor happened to overheat, the reaction rate automatically increased rather than decreased. It was, in short, an accident waiting to happen. Sixteen such RMBK reactors were built in the former Soviet Union. Many of them are still operating. Reactors in the United States and the West, including Japan and South Korea, are designed very differently.

No new U.S. nuclear plants have begun commercial operation since 1996, and most date from the 1970s and 1980s. No nuclear plants were under construction from the 1990s to the time of this writing, and no proposed ones have begun the difficult licensing and approval process. In essence, after Three Mile Island and Chernobyl nuclear energy became a politically toxic subject, and the peak of the cheap-oil fiesta that ran from the 1986 price crash until the attacks of September 11, 2001, allowed the American public and their leaders to stop even thinking about nuclear energy. This situation is apt to change, especially as the United States begins to experience the coming natural gas crunch, which will chiefly affect electric power generation.

The use of so-called breeder reactors could extend the horizon of obtainable electricity from nuclear power further into the future. Breeder reactors use the widely available uranium isotope U-238, together with small amounts of fissionable U-235, to produce a fissionable isotope of plutonium, Pu-239. Plutonium, however, is tremendously dangerous both as a persistent radioactive poison and as a material for bomb-making, and therefore the security requirements for running breeder reactors may be beyond the organizational means of the society we are apt to become in the future, namely one with much weaker central authority, less police power, and reduced financial resources. This is perhaps another way of stating that social stability has been an indirect benefit conferred to us by cheap oil, and in the absence of that oil we can't assume the complex social organization needed to run nuclear energy safely.

In any case, the United States shut down its only prototype breeder reactor and currently has no significant breeder research, development, and demonstration program. Other countries are not doing much better.

Work continues in Japan and in Russia but has ceased in the United Kingdom and France.

Ever since the development of the hydrogen bomb, hopes have been harbored for the development of a commercial *fusion* process that could be used in electric power generation. In fusion, the object is to combine atomic nuclei rather than split them apart—specifically, to bind two hydrogen atoms together to form the element helium. This is the same process that powers the sun, and it produces vast amounts of energy. Human beings replicated this solar fusion process in the development of the hydrogen bomb. Unlike fission, however, we have not yet developed any practical method for harnessing this tremendous force in a way that is controllable. And we are no closer to accomplishing it than we were thirty years ago, during the first OPEC oil crisis, when fusion was one of many alternative fuel miracles promised for the post-petroleum future. A related process called "cold fusion" has been pursued in laboratories sedulously for decades, as methods for turning lead into gold were doggedly pursued by alchemists centuries ago—and so far with similar results.

Perhaps the least obvious aspect of the nuclear conundrum is this: Atomic fission is useful for producing electricity, but most of America's energy needs are for things that electricity can't do very well, if at all. For instance, you can't fly airplanes on electric power from nuclear reactors.[8] The U.S. trucking transport system as currently operated won't run on electricity alone. In the current American mode of living, only about 36 percent of the energy consumed is in the form of electric power generated by one means or another: coal, natural gas, hydro, nuclear. This fraction has remained fairly constant for decades. The rest of our energy comes in the form of burning hydrocarbons. This speaks to the tremendous versatility of petroleum and natural gas. So, as the twenty-first century proceeds, the amount of nuclear-generated electricity is likely to rise, but it

8. The U.S. military actually had a program in the early 1950s to design a nuclear-powered bomber that could stay aloft indefinitely. This was before the successful development of the intercontinental ballistic missile. The idea was to have a fleet of bombers in the air at all times to deter a Soviet atom bomb attack. It became apparent that the amount of shielding necessary to protect the crew from fatal radiation poisoning made the proposed aircraft too heavy to fly.

will not necessarily make up for the losses incurred by fossil fuel depletion (and the costly conflicts over the remaining supplies). It means we can have the lights on at night and refrigerate our food, but without the benefit of artificial fertilizers made out of natural gas, and diesel-powered farm machinery to till the soil at industrial scale, we will have to completely reorganize agriculture. The implication, of course, is that we will have to reorganize virtually everything else in the way we go about our daily lives. But nuclear power may be all that stands between what we identify as civilization and its alternatives.

FIVE

NATURE BITES BACK:

Climate Change, Epidemic Disease, Water Scarcity, Habitat Destruction, and the Dark Side of the Industrial Age

I was at a four-day conference called Pop Tech in the seaside village of Camden, Maine, at the peak of the fall foliage season in October 2003, having a pretty good time at the talks, and enjoying a series of extravagant dinners—one featuring a free oyster raw bar and gratis Grey Goose vodka—not to mention all the lobsters, steaks, and other products of our bountiful cheap-oil economy. Then, on Saturday afternoon, a scientist from the University of Washington, Peter D. Ward, got up in the old-time opera house where the conference was held and did a presentation about the life and death of the planet Earth. Using a series of vivid artist's renderings delivered on PowerPoint, Ward showed us how, hundreds of millions of years hence, all land animals would become extinct, the green forests and grasslands would broil away, the oceans would evaporate, and eventually our beloved planet would be reduced to a pathetic ball of inert lifeless lint—prefatory to being subsumed in the expanded red giant heat cloud of our own dying sun. Few members of the audience had any appetite for the spread of cookies and munchables laid out for the break that followed. Personally, I was so depressed I felt like gargling with razor blades.

The human spirit is remarkably resilient, though. A few hours later, the horror of it all was forgotten and the conference-goers reported to the next supper buffet with their appetites recharged, happy to scarf more lobster and beef medallions and guzzle more liquor, while chatting up new friends about their various hopes and dreams for the continuing story of civilized life here on good old planet Earth, which, it was assumed, had quite a ways to go before any of us needed to worry about its fate, if ever.

Wasn't it John Maynard Keynes who famously remarked to a group of fellow economists dithering about the long-term this and the long-term that: "Gentleman, in the long term we're all dead." Our brains are really not equipped to process events on the geologic scale—at least in reference to how we choose to live, or what we choose to do in the here-and-now. Five hundred million years is a long time, but how about the mad rush of events in just the past 2,000 years starring the human race? Rather action-packed, wouldn't you say? Everything from the Roman Empire to the Twin Towers, with a cast of billions—emperors, slaves, saviors, popes, kings, queens, armies, navies, rabbles, conquest, murder, famine, art, science, revolution, comedy, tragedy, genocide, and Michael Jackson. Enough going on in a mere 2,000 years to divert anyone's attention from the ultimate fate of the earth, you would think. Just reflecting on the events of the twentieth century alone could take your breath away, so why get bent out of shape about the ultimate fate of the earth? Yet I was not soothed by these thoughts, nor by the free eats, and even the liquor failed to lift me up because I couldn't shake the recognition that in the short term we are in pretty serious trouble, too.

There is near unanimity among the scientific community that global warming is happening. There is also a definite consensus emerging that the term "climate change" may be more accurate than "global warming" to describe what we are in for. The mean temperature of the planet is going up. The trend is unmistakable. Average global land temperature was 46.90 degrees Fahrenheit when modern measurements began and had reached 49.20 degrees F in 2003. The rate of change has also increased steadily. The total increase of 2.30 degrees might seem trivial, but has tremendous implications. And the rise in temperature happens to correlate exactly with the upward scale of fossil fuel use since the mid-nineteenth century.

It may not matter anymore whether global warming is or is not a by-product of human activity, or if it just represents the dynamic disequilibrium of what we call "nature." But it happens to coincide with our imminent descent down the slippery slope of oil and gas depletion, so that all the potential discontinuities of that epochal circumstance will be amplified, ramified, reinforced, and torqued by climate change. If global warming is a result of human activity, fossil fuel-based industrialism in particular, then it seems to me the prospects are poor that the human race will be able to do anything about it, because the journey down the oil depletion arc will be much more disorderly than the journey up was. The disruptions and hardships of decelerating industrialism will destabilize governments and societies to the degree that concerted international action—such as the Kyoto protocols or anything like it—will never be carried out. In the chaotic world of diminishing and contested energy resources, there will simply be a mad scramble to use up whatever fossil fuels people can manage to lay their hands on. The very idea that we possess any control over the process seems to me further evidence of the delusion gripping our late-industrial culture—the fatuous certainty that technology will save us from the diminishing returns of technology.

So for the purposes of this book, the relevant question concerning global warming and climate change is not whether human beings caused it or whether we will come up with some snazzy means to arrest it, but simply what the effects are likely to be and what they signify about the way we will live later on in this century.

Surprise!

The news is that abrupt climate change may be normal in the planet's history, or, to state it differently, that the earth's climate is inherently very unstable. The period in which human civilization developed has been, if anything, an anomalous ten-thousand-year epoch of remarkable stability. It is even possible that human agricultural activity during most of this period—that is, before and up to the industrial age—had a paradoxical stabilizing effect on climate, while the profligate burning of fossil fuels during the past two hundred years has restarted the tipping mechanisms that can upset things again. If the current mild climatic interval between ice ages, which we call the Holocene, had not been so extraordinary it would be hard to account for why civilization hadn't burgeoned many millennia earlier. *Homo sapiens* had evolved to pretty much our current level of brainpower fifty thousand years earlier. What postponed the human race's civilizational takeoff?

The answer seems to lie in the Greenland ice cores. Greenland is a protectorate of Denmark; since 1996 the Danish government has sponsored a science project to study the layers of ice in the massive glacier that covers more than 80 percent of the gigantic island (and contains 8 percent of the world's fresh water). The Greenland ice is especially clear and easy to "read." The project involves drilling a five-inch core sample 10,000 feet down through the glacier to the bedrock below. The ice core consists of the accumulated and compressed yearly snowfalls going back more than 100,000 years in discernable layers. The layers record all kinds of evidence about what was happening on the planet in a given year. Ash from the massive explosion of the volcano Krakatoa in 1883 can be found by counting back 122 layers. Various kinds of dust and pollen tell us if drought was occurring or certain plants were flourishing. Even pollution from Roman-period lead smelting can be detected. The layers also contain trapped air bubbles that disclose a record of exactly what the atmosphere was composed of chemically at any given time.[1]

The Greenland ice core samples tell us that the Holocene period has had alternating periods of cooling and warmth within a relatively stable range. The Holocene, at onset, stumbled at first. The retreat from the last ice age was interrupted by a brief but pronounced cooling called the Younger-Dryas period, which lasted about 1,200 years starting around 12,000 B.C. *Homo sapiens* was already established as a species and had been using tools since before the previous ice age. Their numbers and range had been increasing as the glaciers retreated. The Younger-Dryas episode suspended the retreat of the glaciers temporarily and might have provoked the transition from hunter-gatherer culture to agricultural "takeoff." Cooling in the region now called the Middle East would have transformed forested terrain into grassland. Agriculture began as the domestication of cereal grasses. This learning process appears to have taken a few thousand years to complete, during which the climate warmed again and remained unusually benign. The glaciers resumed their retreat. The domestication of animals would have been associated with available grasses and grains. About one-third of the way into the Holocene, around 6000 B.C., this systematic practice of planting and harvesting made permanent settlements and denser populations possible and accelerated the increasing scale of

1. Elizabeth Kolbert, "Ice Memory," *The New Yorker*, January 7, 2002, p. 30.

food production. The greater scale of production eventually led to food surpluses and then to the formation of cities and highly hierarchical societies with a diverse division of labor. Climatic shifts appear to have had a lot to do with the rise and fall of early civilizations organized around intensive food production.

Climatic shifts in more recent times account for historical trends with implications for our present situation. A cooling followed the height of the Roman Empire, bringing, for instance, more rain to England, affecting adversely the health of livestock, and making marginal farming lands submarginal. The Dark Ages were also cool ages. A medieval warming occurred between the ninth and fourteenth centuries; this accounts for historical events such as the settlement of Greenland by the Vikings and the cultivation of vineyards in England. During this period, the human population bloomed with the feudal method of social organization. Toward the end of the medieval warming, the Black Death depleted the European population by as much as one-third and created labor shortages that put an end to feudal social organization. The Renaissance began in Italy as a slight cooling occurred; labor shortages following the Black Death enhanced the value of the individual. A more pronounced dip in temperature produced the Little Ice Age in Europe from the 1500s into the mid-1800s—as chronicled, for instance, by the Dutch landscape painters who show people skating on the frozen canals of Holland, which in the modern era no longer freeze. The average temperature differential between these two periods was only a few degrees, yet the effects were marked. The cooling of the Little Ice Age provoked the deforestation of England and the increased use of coal, and therefore led to inventions for improving coal extraction, namely the steam-powered pump for removing water from coal mines, which soon led to steam-powered railroads and the whole industrial explosion, in which more versatile oil and gas came to replace coal. The brief fossil fuel interval of the past two hundred years has accompanied another warming period, perhaps even stimulated it. And, of course, the amenity of oil has permitted a twentieth-century human population boom like nothing ever seen before. These historical temperature fluctuations, however, may have been minor compared to what we are facing now, especially in light of our suddenly depleting oil and gas supplies.

Stepping back to view the larger-scale picture is sobering. The Greenland ice core record shows that the past 100,000 years have been a climatic roller coaster. It is clear that once climate change begins, it can occur very erratically, a kind of "speed wobble" that ends in a crash. The climax of the last ice age, for example, was 21,000 years ago, when glaciers extended as far south as what is now Connecticut. The transition from the last ice age into the present Holocene was intensely wobbly, including the Younger-Dryas episode. As Elizabeth Kolbert reports: "The temperature did not rise slowly or even steadily; instead the climate flipped several times from temperate conditions back into those of the ice age, and back again. Around fifteen thousand years ago, Greenland abruptly warmed by sixteen degrees in fifteen years or less. In one particularly traumatic episode some twelve thousand years ago, the mean temperature in Greenland shot up by fifteen degrees in a single decade."

It appears that the earth has gone in and out of ice ages on a fairly regular cyclical basis for at least one million years, though the individual cycles show idiosyncrasies of their own. The past interglacial warm period that seems most to resemble the present Holocene is the Eemian, running approximately 130,000 to 110,000 years ago (the point at which the Greenland ice record ends at bedrock). The complete transition from the Eemian warm period to the ice age that followed took no more than 400 years. As the cold grew more severe, the earth's climate also became drier. Water evaporated less effectively from the ocean at colder temperatures and rainfall on land decreased, though ice accumulated from the poles downward. Forests all over the world gave way to drier grasslands and deserts. A slight warming occurred about 60,000 years ago, and then at 30,000 years another cycle of intense cooling and glaciation occurred, which peaked about 21,000 years ago. Around 14,000 years ago there was a rapid global warming and moistening, perhaps occurring within the space of only a few years or decades. The planet was then well on its way to the present Holocene interglacial period—if that's what it is.

We may now be entering a climate speed wobble, which is being aggravated by mankind's release of heat-trapping carbon dioxide. Carbon dioxide (CO_2) is only one of several so-called greenhouse gases that tend to trap heat in the earth's atmosphere, or prevent it from radiating into space, and it is the least effective by far of the three main culprits. The most effective heat-trapping agent is water vapor. The next one is methane, approximately twenty times more effective as an earth insulator than CO_2. Methane is a by-product of agriculture (especially from rice paddies or released in the excrement of domestic animals) or produced by decay in swamps (methane is also known as "swamp gas") or by the thawing of organic matter in warming tundra. That said, however, carbon dioxide is certainly an effective greenhouse gas, and the amount of carbon dioxide as a percentage of the atmosphere today has not been so high since the days of the dinosaurs. By putting large quantities of greenhouse gases into the atmosphere, humans are exerting pressure on an inherently unstable climate system that might produce a drastic change without much prior warning.

Apart from the levels of greenhouse gases, the causes of climate change probably include periodic variations in the sun's radiation, variations in the earth's orbit, volcanic activity and particulate matter in the atmosphere, and the extremely complex interplay of the oceans and the atmosphere that circulates heat and cold around the planet. The last can take on the characteristics of a positive feedback loop, and in the short term presents the greatest threats to our very delicate project of civilization.

The Gulf Stream "Switch"

The Gulf Stream is an idiosyncratic feature of hemispheric topography that allows the transfer of huge amounts of tropical warmth from the Gulf of Mexico to northern Europe, making that region superbly hospitable to human settlement. Without the Gulf Stream, Britain, France, the Low Countries, and Scandinavia would have a climate like Labrador's, colder by twenty degrees Fahrenheit in annual mean. The Gulf Stream has been likened to an oceanic conveyor belt. The force of the warm water flowing north has been described as equal to the volume of seventy-five Amazon rivers. This powerful churn is also thought to affect the behavior of currents as far away as the Pacific and Indian oceans.

The conveyor belt works because the chemical and physical properties of the water at each end differ slightly. The warm waters from the Gulf Stream move northward, propelled by the earth's rotation. These waters are already unusually salty, because in the warmer latitudes surface water evaporates more readily. By the time the north-flowing Gulf Stream bumps up against Greenland and Iceland, the water has shed much of its original warmth. The cooler water is now saltier, denser, and heavier. It sinks with enough velocity to push back south again as a deep-water current. Meanwhile, back in the Gulf of Mexico, more warm surface water begins its journey north on the conveyor belt. The process is continuous.

There is evidence that global warming today is causing ice sheets in the far north to melt, sending large quantities of freshwater into the North Atlantic where the cooling, dense, salt-heavy waters make their plunge to return south. The fresh melt water dilutes the cold salt water, making it less dense, less heavy. It is feared that this will impede its sinking, and thus slow the whole conveyor belt of the Gulf Stream. There is, in fact, evidence that the cold water flow has weakened by 20 percent in recent years.[2]

A crucial aspect of the problem is that a seemingly slight salinity differential between the warm and cold currents—one part per thousand—is all that it takes to drive this gigantic natural mega-machine of the Gulf Stream. It might not take much in the way of temperature change to alter it either, even to push it over a threshold that would abort the circulation altogether, shut it off. A prevailing theory holds that the Younger-Dryas episode was caused by an earlier shutdown of the Gulf Stream, in response to a sudden influx of fresh water from deglaciation in North America—perhaps the sudden catastrophic breaching of an inland freshwater lake system that accumulated during the retreat of the glaciers. The Younger-Dryas cooling would have halted further glacial melt for a millennium, allowing salinity to build up again in the northerly flow of the Gulf Stream and eventually restarting the conveyor belt, which allowed the Holocene to resume.

The mechanism for this paradoxical flip-flop process may now be evident. Warming occurs until the Arctic ice melt "switch" turns off the Gulf Stream, which then induces cooling around the Arctic region, which eventually readjusts the salinity of the northern waters and turns on the cycle again. It may not matter that much whether the initial

2. *U.S. News and World Report*, April 1, 2002, p. 65.

warming is caused by gases released naturally by swamps and animals or by American commuters driving Ford Explorers. One of the inferences of this theory, however, is that we don't really know whether the current episode of global warming may, in fact, only be the prelude to another major cooling. What we face, actually, may be a one-two punch: a radical warming episode disruptive in its own right followed by the onset of an ice age.

If historical patterns are any measure, the Holocene is about due to end after a 10,000-year run. Perhaps the most amazing thing of all, as Kolbert points out, is that "the only period in the climate record as stable as our own *is* our own. And it seems even more improbable that climatologists should make the discovery that we are living in this period of exceptional stability at the very moment when, by their own calculations, it is likely nearing an end."

For the moment, however, at least for the decades just ahead, it appears that we are in for the warming phase of that one-two punch. Combine that with the roundhouse blow of our depleting oil and gas reserves and that might be enough to knock our fine-tuned civilizations to their knees long before the next ice age gets under way.

Global warming has accelerated during the past twenty years. While the trend toward warmer temperatures has been uneven over the past century, the trend since 1976 is roughly three times that for the whole period. The ten hottest years in the 143-year-old global temperature record have now all been since 1990, with the three hottest being 1998, 2001, and 2002. Global average land and sea surface temperatures in May 2003 were the second highest since records began in 1880. Considering land temperatures only, May 2003 was the warmest on record. Climate models project that the earth will warm by 2.5 to 10.8 degrees Fahrenheit between 2000 and 2100, with most land areas warming more than the global average. Eleven of the last twelve years have been the hottest ever recorded since scientific instruments came into use. The August 2003 heat wave in Europe caused an estimated 30,000 deaths, including more than 13,000 in France alone. Temperatures reached 104 degrees F in Paris and over 100 degrees in England (a record). Temperatures in the hardest-hit areas of middle and southern France averaged eighteen degrees higher than normal. Parts of Italy suffered comparably. In London, trains were shut down over fears that tracks would buckle in the heat, while in Scotland the high temperatures, combined with falling water levels in rivers and streams, threatened the spawning and survival of salmon. Switzerland reported its hottest June. Throughout France, Spain, Portugal, Italy, Poland, and the Balkans, the intense heat and dry conditions sparked devastating forest fires. The heat wave was accompanied by an unprecedented drought that devastated crops. France's wheat loss was 20 percent, England's 12 percent, and Ukraine's a staggering 80 percent.

In the previous year, 2002, Central Europe had been afflicted with unprecedented floods, called the worst in five hundred years and resulting in more than $15 billion in damage. In Dresden the Elbe River reached thirty-one feet above flood stage, the highest level since records began in the sixteenth century. Prague was heavily damaged as floodwaters caused the collapse of many ancient buildings and 200,000 residents were forced to evacuate their homes. Floods in China and Southeast Asia the same year killed even more people than the floods in Europe had. The 2003 tornado season in the United States was the most active ever recorded, with a record-breaking 562 twisters in May alone. (The previous record was 399 in June 1992.) The 2004 hurricane season in the United States was especially severe, with three major storms striking Florida in a six-week period and more than $20 billion in damage incurred there, not to mention devastation in several Caribbean nations.

The Intergovernmental Panel on Climate Change (IPCC) predicts that severe storms, extreme floods, and droughts will become more pronounced as global warming advances.[3] In general, we are in for much more unstable weather in the decades ahead. The effects will be complex and vary considerably from place to place. For instance, while Europe broiled in the summer of 2003, the northeast United States breezed through an eerily cool summer, with few days over 90 degrees all season. An altered jet stream pattern prevented southerly air from

3. Recognizing the problem of potential global climate change, the World Meteorological Organization (WMO) and the United Nations Environment Program (UNEP) established the Intergovernmental Panel on Climate Change (IPCC) in 1988. It is open to all members of the UN and WMO.

penetrating the Northeast. In fact, on one of the very few torrid days that whole summer, August 14, a surge in demand for air conditioning took down the electric grid in the Northeast.

The changes now occurring are sometimes surprising. For instance, the actual warming trend shows up much more in increased nighttime temperatures than in mean daily temperatures. Likewise, the U.S. National Climatic Data Center has determined that the minimum temperatures—that is, the day's low temperatures—have been rising more than the highs. One result this translates into is later frosts in the Northern Hemisphere. In the northeast United States, for example, the frost-free season now begins an average of eleven days earlier than it did during the 1950s. This might seem beneficial for growing crops but it is also problematic, as the rapid change destabilizes existing ecologies, harming established species and creating favorable conditions for invading newcomers that will, in turn, cause additional ecological changes, many of them damaging.

In another paradox, as the global mean temperature has gone up, winters have become more snowy in the high latitudes of the Northern Hemisphere—because greater heat at lower latitudes is causing greater evaporation of moisture. This is another bit of evidence suggesting that the current warming may actually be a prelude to another ice age. In the middle latitudes of the northern United States and Canada, the ratio of winter snow to rain has decreased, while overall winter precipitation has increased. In spite of the increased winter moisture, soils in North America are expected to become drier in the decades ahead, because warmer temperatures will more effectively boost levels of evaporation and transpiration. Cloud cover, for instance, can alter the calculus behind these effects, however, and indeed some parts of Russia are evidently seeing increased levels of soil moisture.

Climate and the Food Supply

There are two salient issues concerning agriculture and climate change in the years directly ahead. First is the fact that the effects will differ from one region to another. The second is that the disturbances of climate change will intersect with the decline of oil and natural gas resources, which have allowed human beings to practice an extraordinarily productive kind of industrial agriculture for about a century and enabled a fantastic meta-growth of human population. Climate change is going to combine with the termination of oil-and-gas-based farming to very negatively affect the world's food supply. A lot of people will go hungry in the decades ahead and many of them will die.

Where the current phase of global warming is concerned, we really do not know yet what the geographic distribution of warming will be—that is, which places will get a lot warmer, a little warmer, hardly warmer at all—and what the effect will be on rainfall patterns or the ability of the soils in a given region to retain moisture. Eurasia may benefit and the Great Plains of North America may suffer, or vice versa. The growing season in both places may get longer, but the benefit of that may be canceled out by drought in both regions.

We also don't know what the response of crop plants will be to significantly increased carbon dioxide levels in the atmosphere. Wheat, rice, and soybeans tend to accommodate higher levels of CO_2. They photosynthesize more carbohydrates. Corn and sugarcane are not so happy with increased CO_2. However, excessive heat could easily vitiate any theoretical benefit of increased CO_2. None of these crops thrives in drought conditions. Theoretically, global warming and a longer growing season might allow grains to be cultivated at higher latitudes in Canada and Russia. But a shortage of natural gas as the basic feedstock for fertilizer would make the benefit moot, except at the margins—say, where previously uninhabitable lands might be settled and farmed without oil and gas inputs; in other words, postindustrial subsistence farming.

Falling water tables will also make grain production problematic, especially in areas of the United States currently dependent on "fossil" water from the Ogallala aquifer, including parts of Nebraska, Kansas, Oklahoma, Texas, Colorado, and New Mexico. The Ogallala has been depleting at an accelerating rate for decades, with annual use way outpacing replenishment. Thirty-three percent of all crops produced in the United States are grown on irrigated acres. More severe heat waves will strain peak irrigation demands and aggravate competition between urban areas and farms. Also, increased heat and evaporation on irrigated lands will intensify the accumulation of salts in the soil—an age-old hazard of societies that depend on irrigation over a long period of time.

Under the North China Plain, which produces half of China's wheat and a third of its corn, water tables are falling by three to ten feet per year. Along with rising temperatures and the loss of cropland to nonfarm uses this trend is shrinking the Chinese grain harvest. China raised its grain output from 90 million tons in 1950 to 392 million tons in 1998. Since 1998, though, China's production appears to have peaked, dropping by 66 million tons, or 17 percent.

We do know exactly which are the leading grain-producing regions of the world now, and we can state that the world is now not quite getting by on what is produced globally. There are plenty of people already starving. There is little likelihood that all of the current major grain-growing regions will either benefit or be unaffected by global warming. There will be winners and losers, but on the whole there will be more losers. The human race is going to suffer. Recent history offers a preview. The year 2002 was the climax of a sustained multiyear drought in the United States that resulted in the worst wheat crop in twenty-eight years. Canada's "breadbasket" was affected, too. Australia's wheat production also tanked that year due to drought. In the 2002–2003 market year 99.7 million tons of wheat were traded worldwide, down from about 107.4 million tons the previous year. World population growth did not decline that year. While exports from the United States, Canada, Argentina, Australia, and the EU, the five largest exporters, were down nearly 25 percent from the previous year, the 2002 harvest in Russia, Ukraine, and Kazakhstan was the largest since the breakup of the USSR. The following year European production got hammered by extreme heat and drought, while U.S. production recovered. The overall trend in world cereal grain production has turned down in recent years.

The so-called Green Revolution of the late twentieth century increased world grain production by 250 percent. The increase was almost entirely attributable to fossil fuel inputs: fertilizers made out of natural gas, pesticides made from oil, and irrigation powered by hydrocarbons—with a little kicker from plant genetics. In the year 2000, the wellhead price of natural gas skyrocketed 400 percent, the sharpest energy price increase the nation had ever seen, outdoing even the oil spikes of the

1970s. When U.S. natural gas depletion accelerated in 2002–2003, many American fertilizer operations closed down or moved overseas. Future prospects range from a best case of much higher prices for fertilizers to a worst case of desperate shortages of methane-derived fertilizer. Global warming can only make matters worse.

Warmer conditions are favorable for the proliferation of insect pests and plant diseases. Longer growing seasons will enable insects such as grasshoppers to complete a greater number of reproductive cycles during the spring, summer, and autumn. Larvae of other bugs will winter over more comfortably. Any upset of an established ecologic balance is an invitation to previously unwelcome organisms. Even during the most stable past decades, parts of the United States suffered from boll weevil, locust, and gypsy moth infestations. Climate change will likely increase both the intensity and diversity of invading agricultural pests and diseases. A very long list of new plant diseases is just lately emerging, from barley stripe rust to eastern filbert blight to silver scurf (potatoes) and verticillium wilt (strawberries), which could make farming very difficult and disappointing in the future, whatever way it is practiced.

Unfortunately all the mathematical computer models created to puzzle out the effects of global warming on agriculture seem to be based on the idea that climate change alone is the only major variable. They do not take into account factors such as the permanent and growing shortage of natural gas in the United States, or the global oil peak story, or the vulnerability of global markets to social turbulence, or the probability of armed conflict over oil, water, or territory, and the interplay among them all. From what we have seen of weather in the United States over the past ten years, one could already conclude that global warming will not be so benign here, especially in the deep continental interior. We have had a decade of increasingly extreme weather, especially heat and drought. The Great Plains of the United States may be among the losers in the global warming derby. The American public is already faced with the task of radically reorganizing the way farming is done, as the industrial model fades into irrelevance and the giant combines stop running on the megafields of the big corporate farms. The conditions ahead may be such that America will be challenged to produce enough food for its own domestic needs, never mind exporting to the starving masses in other nations. Overall, in the Long Emergency ahead, human existence everywhere will have to become more local in every way, and the various winners and losers will not be so able to compensate one another.

Environmental Destruction

The damage to global ecologies by human activity accelerated rapidly with the onset of industrialism. The twentieth century, with its oil-nurtured bloom of human population, was especially harsh. Everywhere, biological complexity was compromised or reduced to monoculture. Habitats were wrecked. Species were exterminated. Terrain and water were poisoned. The amount of asphalt paving alone in the United States represents an ecological insult beyond calculation. These man-made environmental catastrophes will combine with and be reinforced by the new problems of climate change in the Long Emergency.

Among the more melodramatic negative advertisements for what's coming is the news that sea levels are rising and are likely to rise a lot more in this century. Places such as Bangladesh, the Netherlands, much of Florida, the Chesapeake Bay lowlands, New Orleans, and scores of Pacific islands could vanish underwater in the coming century. Harbor towns all over the world could be damaged or submerged. The worldwide ice meltdown is impressive. The Arctic has been warming dramatically and virtually all glaciers, ice caps, and ice sheets are melting. Sea ice is also retreating and thinning. The Alps have lost half of their glacial mass since 1850, with the loss accelerating during the past twenty years. Measurements by aircraft using global positioning satellites and laser altimeters show that the 5,000-square-kilometer Malaspina glacier in Alaska is losing nearly a meter of thickness per year—the equivalent of three cubic kilometers of water. Antarctic ice sheets are breaking up at an unprecedented rate. Between January and March 2002, two-thirds of the giant Larson B ice shelf collapsed, greatly altering the salinity of the surrounding seas and killing off plankton and krill that were the basis of the region's wild food chain. The 700-foot-thick Larson had been in place for an estimated 12,000 years, and had lost 3,420 square miles since 1997 because of warming temperatures in the region. Only about 40 percent of the shelf remains. The Ross ice shelf, Antarctica's largest at 332,000 square miles (about the size of France), may be next. Icebergs the size of Rhode Island and Manhattan are breaking off at a rate too great to be recorded. Generally, the temperature rise in the southern oceans has been greater than elsewhere around the globe.

According to the IPCC, sea levels rose by ten to twenty centimeters during the twentieth century, and are currently rising by about two millimeters a year, which is at the upper range of the rate of rise for the last century. With global warming accelerating, this is apt to increase. The generally accepted prediction is that sea levels will rise during the twenty-first century by about fifty centimeters, or a little under two feet, though some scientists predict a full meter. Roughly one-sixth of the people in the world live in coastal zones within one meter of sea level.

This is the kind of outside context problem so alien to contemporary experience that the public and its leaders can really find no way to process the information and figure out what to do about it—and for the excellent reason that it is not a problem with a direct solution. It is more a condition without a remedy. If the major shipping ports of London, Bombay, Yokohama, Norfolk, San Pedro, and so on, end up being submerged, humankind will just have to work around it. The disruptions to world trade might be epochal, gigantic, ultimately tragic. It seems obvious that the human race will simply have to adjust, even if that means adjusting to a new reality of severely lower expectations in living standards, comfort, and amenity. In the meantime, however, there is virtually no public discussion of this prospect in the United States now, no talk about making other arrangements. When the time comes, I suppose, many Americans will just have to move to higher ground.

The vast majority of the earth's surface consists of water, yet only 3 percent of that is fresh water. The World Bank has famously declared, "The wars of the twenty-first century will be fought over water." The United Nations has identified three hundred zones around the world that will be the sites of conflicts over water in the years ahead. The great aquifers of North America, China, and India are all depleting rapidly due to aggressive irrigation—up to 70 percent of all cropland in China, for instance—made possible by cheap fuel. This is one way we can understand the direct conversion of oil into food. The rapidly diminishing supplies of fresh water, especially in the heavily populated third world, also exacerbate sanitation catastrophes, and prepare the stage for epidemic disease. More than two million people worldwide die every year from contaminated water. In the maquiladora zones of Mexico today, water is so scarce that babies and children drink Coca-Cola instead.

Harvard biologist Edward O. Wilson warns that China's current program to mitigate huge population increases with gigantic water projects may have dire consequences.[4] Irrigation and other withdrawals have already depleted the Yellow River, which, starting in 1972, has run bone-dry part of the year in Shandong province, where one-fifth of China's wheat and one-seventh of its corn is produced. In 1997, the river stopped flowing for a record 226 days. The groundwater levels of the northern China plains have plummeted. The water table in major grain-producing areas is falling at the rate of five feet a year. Of China's 617 cities, 300 already face water shortages. Of China's approximately 23,000 miles of major rivers, 80 percent no longer support fish life.

The Xiaolangdi dam project now under way along the Yellow River in north China is exceeded in size only by the Three Gorges Dam on the Yangtze in south China. In addition, the Chinese government intends to siphon water from the Yangtze—which has not yet run dry—and send it over by a canal system to the Yellow River and Beijing, respectively. When it is running, the Yellow River is already one of the most particle-laden in the world. Because of that, it is estimated that the Xiaolangdi dam would silt up within thirty years of completion. The $58 billion project is reminiscent of another centrally planned mega-project that ended in grief: the Soviet Union's scheme to drain the Aral Sea to irrigate gigantic cotton farms in Kazakhstan. The project turned one of the world's largest inland bodies of fresh water into salty desert.

The potential for calamity in China is therefore huge as it skirts a range of forces presented by the Long Emergency, any one of which, or some combination, could send it reeling over its tipping point: the effects of global climate change, competition for oil, extremes of pollution, disease, and war, either with its neighbors or internally. Despite the current veneer of prosperity and stability, China has tremendous potential

4. Edward O. Wilson, "The Bottleneck," *Scientific American*, February 2002.

for political chaos. As Wilson fearlessly points out, the pressure on China's agriculture and water resources is intensified by the predicament shared by many countries: runaway population growth. Population growth rates may be mitigated somewhat from culture to culture by economic advance (which tends to lower reproductive rates by channeling women into the workplace), but economic development produces other not-so-benign consequences. Developing nations invariably increase their energy use. More cars are used, more electricity generated, more greenhouse emissions sent into the atmosphere. In the Long Emergency, to borrow a remark from author James Flink, "there will only be two types of nations: the over-developed and those which will never develop."[5] China may represent an amalgamation of those two conditions in one nation-state.

The looming worldwide crisis over freshwater beats another path to a crisis of food production. "We have created a food bubble economy," writes ecological ethicist Bill McKibben. "Just as we have built our industrial economy on cheap oil, so have we managed to artificially inflate food production by an unsustainable reliance on underground water. The pumping of groundwater has generated tremendous crop yields . . . but when the water starts to run dry, that free ride is over and farmers will have to return to growing what they can with the water that falls on their regions."[6] A reasonable inference is that some of these regions will not just sink quietly into hunger and desperation but that wars will be fought for survival. Certainly, famine and want will drive huge movements of population in the opening stages of the Long Emergency. Climate change and environmental degradation threaten the ability of many of the poor to escape poverty and increase greatly the prospects for political instability and military strife. Nothing is more fundamental in history than wars over resources. As regions are stressed by food and water scarcity, by droughts, floods, and other weather-related calamities that destroy homes and communities, those that retain military capability will be tempted to compete for survival by means of aggression. Other poor and desperate regions will

5. James Flink, *The Automobile Age*, Cambridge, MA: MIT Press, 1988.

6. Bill McKibben, "Our Thirsty Future," *The New York Review of Books*, September 25, 2003.

see their populations migrating. Some may see both anarchy and out-migration. The worldwide proliferation of small arms, submachine guns, shoulder-launched rockets, and portable missiles has given even the most wretchedly impoverished people the ability to challenge the best-equipped armies in the world, to engage, encumber and even defeat them.

Like China, the United States is divided roughly in half between wet and dry. Though the human population of the United States is proportionately much smaller than China's, the amount of effort America has expended on manipulating habitats and altering terrain is as impressive in its own way as China's birthrate. Especially significant is the stupendous amount of paving laid down in the United States during the past hundred years. It prevents rain from being absorbed as groundwater and sends it instead into rivers, and ultimately into the ocean. The effect of this is the inability of water tables and wetlands to recharge and the diminishing ability of the terrain to support life. In the United States, only 2 percent of the country's rivers and wetlands remain free-flowing and undeveloped. As a result, the country has lost more than half of its wetlands.

The U.S. average of 1,300 gallons of water per day, per citizen, is the highest use rate in the world, and some sixty times the average for many third world nations. Low density suburban sprawl is the fastest-growing sector of water use in the United States now. Both suburban Atlanta and suburban Denver are virtually tapped out, unable to increase their water supply under any circumstances. Dallas and San Antonio are not far behind. Las Vegas hallucinates its future water supply, and southern California is at the mercy of the Sierra and Rocky Mountain snowpacks, which in recent years have shown alarming declines. Global warming implies that a greater proportion of the annual precipitation in the American west will fall as rain rather than snow. The snowpack acts as a storage reservoir, releasing water in summer time when demand peaks. If that precipitation falls as rain instead, it will flow into rivers and streams and run off into the Pacific Ocean at a time of year when demand is lower. The result will be summer crises in both water and power generation. A joint study by a consortium of U.S. agencies and institutes projected that over the first half of the twenty-first century a one-third drop in reservoir levels along the Colorado River would cut hydropower generation by as much as 40 percent. The same study also predicted reduced flows in the Sacramento River and the Columbia River.[7]

Much of Florida is barely a few feet above sea level. It is not necessary for the tides to wash over Dade County, or for a category-five hurricane to strike, for "normal" American life to be endangered there. Most Floridians live within ten miles of the coast. On the Atlantic side, they depend on the Biscayne aquifer for their fresh water. More than 90 percent of Florida's population depends on groundwater as the source of drinking water for public and private wells. If ocean levels rise even marginally, seawater will invade the Biscayne aquifer and Floridians will have to make other arrangements. At the upper margins of global warming prediction, a reduction of the West Antarctic and Greenland ice sheets similar to past reductions could cause sea level to rise ten or more meters. A sea-level rise of ten meters would flood about 25 percent of the U.S. population, with the major impact being mostly on the people and infrastructures in the Gulf and East Coast states.

The nation currently has about 50,000 separate municipal or county water systems, and to aggravate matters, the existing infrastructure of pipes in most U.S. cities and towns is decrepit. Much of it was originally installed in the early decades of the previous century. Water main breaks run around 238,000 incidents a year.[8] In Atlanta in recent years, the water coming out of the highly stressed Chattahoochee River was so turbid that at times citizens could not see the drains in their filled-up bathtubs. The urgent need to replace this massive infrastructure will confront the reality of a nation entering functional bankruptcy in the Long Emergency. Climate change, competition for water, and polluted water sources will also be exacerbated by failures in the electric grid caused by oil and gas supply disruptions. Even if water is available, localities may lack the power to push it through their treatment plants and municipal pipes.

7. The Scripps Institute, The University of Washington, the U.S. Department of Energy, and the U.S. Geological Survey, as reported by the Associated Press, December 23, 2002.

8. *U.S. News and World Report*, August 12, 2002.

Return of the Grim Reaper

Fifty years of easy living with the miracle of antibiotics was a major contributor to the hubris that gripped the industrial nations in the early twenty-first century. Smallpox was eliminated except in strategic laboratory samples. Measles was conquered. Sexually transmitted diseases that used to leave people maimed and crazy were cured with one visit to the doctor. Many tropical diseases seemed to be on the wane as immunology and pharmacology bolstered widespread progress in sanitation and nutrition. The vanquishing of disease represented a kind of meta-victory by mankind over a much greater set of enemies than the parochial combatants of our geopolitical wars. Indeed, these great advances of medical science against disease took place against the backdrop of war. The United States emerged victorious from the last great world war, having defeated manifest political evil, armed with penicillin and sulfa drugs. The postwar antibiotic miracle contributed to a false sense of security in the public and a sense of adolescent-like omnipotence among leaders in science, business, and politics.

By the 1960s, the dream of defeating death itself began to seem as plausible as the dream of landing on the moon. Even our architecture reflected it. The sterile boxes of corporate headquarters in New York City, the retort-like museums going up in our cities, looked like nothing so much as pieces of laboratory equipment, tributes to our prowess in science. In 1967, the surgeon-general of the United States, William H. Stewart, famously declared that it was "time to close the book on infectious diseases, declare the war against pestilence won, and shift national resources to such chronic problems as heart disease."

Then along came AIDS and a horrible host of other new illnesses, and a resurgent gang of old adversaries such as tuberculosis, malaria, and staphylococcus, now resistant to our wonder drugs, and that glow of hubris began to darken into the sick, greenish aura of postmodern ironic angst—as if to say, "Wasn't all that postwar self-confidence a joke." Even the milestone play *Angels in America* regarded AIDS as a cosmic gag as much as a tragedy. The few decades of medical triumphalism, of polio shots and surefire cures for the kinds of infections that routinely killed our ancestors, will someday seem like a golden age of good health. In the Long

Emergency, mortality will return with a vengeance. In fact, falling life spans will characterize this period as much as anything.

There are four basic categories of diseases that pose different kinds of threat to the American public: (1) the new diseases, including AIDS, SARS, bovine spongiform encephalopathy ("mad cow disease"), and "designer" bugs developed in labs; (2) the old standard diseases with developed immunity to antimicrobial drugs; (3) invading vector-borne exotics moving into new territory, such as dengue fever, malaria, West Nile virus, and Lyme disease; and (4) viral epidemic influenzas.[9] Some diseases apply in more than one category.

A major cause of this disease activity is the massive and effortless transportation of people and goods around the world, and at the bottom of that, of course, has been cheap oil. Cheap oil makes it possible for an airplane to carry live Asian mosquitoes to the Caribbean, where those mosquitoes have unleashed dengue fever, or for a person infected with SARS to travel from China to Toronto in a day's time, or for someone to carry AIDS from Johannesburg to Atlanta. As the Long Emergency proceeds, and globalism winds down, this kind of travel and traffic will decrease, but much of the damage has already been done. West Nile virus and dengue fever are already established in places where they had not previously been. They are probably there to stay. If anything, global warming will now likely extend their range.

AIDS has already made the crucial genetic leap from being a disease established in apes in a wild backwater of Africa to now being a public health catastrophe on every continent of the globe, with no cure in sight. In 2004, roughly 40 million people worldwide were infected with the human immunodeficiency virus (HIV) that causes AIDS (acquired immunodeficiency syndrome). Among them are 2.5 million children. AIDS is the leading cause of death in Africa and the fourth leading cause worldwide. The infection is spreading at a doubling time of 5.7 years. AIDS is especially problematic because it takes such a long time to actually kill its victims. Typically, symptoms of immune system failure will take seven to ten years to present from the time of actual infection with HIV. That

9. A disease vector is an agent of transmission—for instance, a particular species of mosquito associated with a specific illness.

affords tremendous opportunity for the organism to spread. Even if by some miracle all transmission of HIV stopped tomorrow, many people would still become ill for decades to come. We are only at the beginning of the impact of AIDS, certainly in Africa and Asia.

In the United States, deaths from AIDS had slowed due to anti-retroviral drugs. But by 2004 roughly three-quarters of newly infected HIV cases were showing resistance to one or more of these drugs. The virus has an impressive ability to adapt to new conditions. Affluent America was exceptional, in that so many HIV cases could even be treated with expensive courses of drugs, which typically cost thousands of dollars a year. Since the drug regimens were complex, there were also significant numbers of patients who did not take their pills correctly, which gave the virus opportunities to adapt and become resistant. In parts of the world with the highest rates of infection—sub-Saharan Africa, for instance—the vast majority of HIV-infected persons are so poor, or so cut off from public health services, that drug therapy is out of the question.

Worldwide AIDS cases were expected to more than double by 2010 to about 80 million, with the fastest growth in China, India, Indochina, Indonesia, Russia, eastern Europe, the Caribbean, and Brazil. In Africa, where the disease has burned through the population the longest and most severely, a dozen nations have infection rates of over 10 percent, with overall numbers of new cases still continuing to increase. In Botswana and Swaziland—two former tribal protectorates within South Africa that are now independent states—the infection rate is 39 percent. The social devastation accompanying the epidemic in Africa is a preview of what is liable to happen in other parts of the world as the Long Emergency gathers momentum. The trajectory of the disease in Africa has left a series of countries in brokenness and anarchy. Sub-Saharan Africa may be significantly depopulated by 2025.

To date AIDS has spread primarily by means of sexual contact or by the sharing of unsterilized needles by drug users. In China, the disease is believed to have reached takeoff through poorly run blood collection programs via the reuse of needles among poor peasants who routinely sold their blood for subsistence income. But in that country the means of transmission is now increasingly a result of sexual practices and drug injection, as in the rest of the world. The Chinese government has aggressively suppressed information about the epidemic for reasons that are both political and cultural. The rapid spread of AIDS will heavily tax the economies of large and dynamic countries such as China and India, whereas in Russia the public health system had already substantially broken down with the demise of the Soviet bureaucracy. Russia, with its decrepit infrastructure, imploded industrial economy, tattered social safety net, and demoralized citizenry, is the prototype for the fate of industrial societies of the Long Emergency.

There are no signs that the AIDS epidemic is leveling off. With its doubling period of 5.7 years, AIDS is on a collision course with the coming disruptions of global oil supplies to perfectly synergize into widespread social turbulence. As the struggle over the remaining oil and gas intensifies, larger numbers of economic losers will be created, and those economic losers will be underfed, ill-housed, poorly doctored, badly informed, badly behaved, and subject to plummeting life expectancies.

There is also reason to suppose that the mutation of the resourceful HIV virus is not over. It could find new ways to spread other than its current favored methods. A model for this is modern syphilis, which is believed to have mutated out of the spirochete responsible for yaws, and presents symptoms almost indistinguishable from its precursor. Yaws, a form of treponematosis, originated in Africa and probably spread into the Mediterranean during the early Roman period. Yaws enters the body by skin-to-skin contact. In ancient times it was probably confused with leprosy. As better woolen clothing developed during the Middle Ages, the spirochete causing yaws faced a crisis of survival and, by mutation, sought a surer pathway into the human body, which it accomplished by genital congress. One great fear lurking in the epidemiology of AIDS now is that the HIV virus could find a means to spread by aerosol means—exposure of mucous membranes to coughing and sneezing. In this event, the consequences for the human race could be fateful to the extreme.

Influenza is an equally potent threat to populations swollen by the twentieth century's fossil fuel hypertrophy, though its mode of attack is much different from AIDS. For one thing, flu can kill in a few days after infection and it does not rely on any particular form of human behavior to spread. An epidemic requires only large cosmopolitan populations to take off. The flu virus originated in wild aquatic birds, has spread and mutated in domestic fowl, and tends to jump species upward, first to domestic swine, which serve as transfer breeding stations for the virus, and then to humans. Pigs seem to act like living laboratories where bird and mammal viruses can get together and share RNA segments to create new strains of flu virus. Where human populations swell and more people mingle with chickens, ducks, and pigs, the prospects increase dramatically for new brands of flu. This is the case particularly in China, where a peasant population of a billion lives in close quarters with their animals.

Flu spreads easily by coughing, sneezing, or skin contact. It still travels efficiently in wild birds, which swap the disease back and forth with their domestic cousins. Flu mutates continually, like the figures spinning on a slot machine, and every now and then—in about eighty-year cycles—hits a jackpot to produce new strains that are violently destructive to human life. The garden-variety flu is a disease that causes fever, joint pains, chills, sore throat, headache, and exhaustion and can be life-threatening to old people and children. But the occasional super-flu escalates beyond those common symptoms to severe pneumonia, toxic shock, and organ failure, even in the young and healthy. In fact, the preponderance of deaths worldwide from the 1918 flu were among people age twenty to forty, a circumstance that is still poorly understood.

The infamous 1918 influenza, which spun around the planet in the last year of World War I, ended up killing more people than the Great War itself. Though it was called the "Spanish" flu, it most likely originated in Asia. In America, the first outbreak occurred on a military base in Kansas in the vicinity of a pig farm. The 1918 pandemic is believed to have killed up to 40 million people worldwide, including 675,000 Americans. Many more got sick but survived. The 1918 flu killed more soldiers engaged in the war than the combat of the trenches. More people worldwide died of that influenza in a single year than in four years of the bubonic plague from 1347 to 1351.

The 1918 pandemic affected everyone. With one-quarter of the United States and one-fifth of the world infected at its height, it was nearly impossible to escape from the illness. Even President Woodrow Wilson suffered from the flu in early 1919 while negotiating the crucial treaty of Versailles. Those who were lucky enough to avoid infection had to deal with awkward public health measures to restrain the spread of the disease. Gauze masks had to be worn in public. Stores could not hold sales; funerals were limited to fifteen minutes. Some towns required a signed certificate to enter and railroads would not accept passengers without them. Those who ignored the flu ordinances had to pay steep fines enforced by public health officers. Bodies piled up as the massive deaths of the epidemic continued. Besides the lack of health care workers and medical supplies, there was a shortage of coffins, morticians, and gravediggers.

The world is overdue for a new outbreak of supervirus on the order of the 1918 flu. One like it may have been only narrowly averted. In May 1997 Hong Kong officials reported the death of a three-year-old boy from complications of influenza. In August, authorities identified the strain of influenza virus isolated from the boy as H5N1. This flu previously had been known to exist in shorebirds and to occasionally infect chickens, but this was the first time a human being had been found to be infected with this particular influenza strain. What made the case all the more strange was that the virus had jumped directly from bird to human. Over the next four months, the virus turned up in twenty additional human cases, six of which resulted in death. Three days after Christmas, worried officials began a massive slaughter of all the poultry in Hong Kong—1.4 million chickens, ducks, quail, and other birds.

It was assumed that infected persons had caught the illness directly from live poultry. What frightened public health officials was the prospect that a flu virus like avian H5N1, having demonstrated its ability to infect humans, might jump species to transmit itself from human to human, not just bird to human. That would have been the takeoff point of a very severe human-to-human influenza that could kill a third of its victims and spread with lightning speed around a hyperconnected world.

Since the Hong Kong incident, there have been scores of bird flu outbreaks, most of them in China and other Asian nations, and some in the United States, where battery breeding of poultry takes place at fantastic scales of production. None of these so far has been as virulent as the 1997 Hong Kong outbreak, but that is only a matter of luck. All of them have resulted in the mass slaughter of infected birds, so the damage has been mostly economic.

The huge rise in world population and relative remission of global warfare in the decades since 1945 has also seen a tremendous increase in the factory farming of animals both in sheer numbers and scale of operation. This has led to many unhappy consequences, some of them rather arcane. For instance, when hurricanes Floyd and Irene successively struck North Carolina in 1999, the damage they caused was due not to high winds as much as flooding from torrential rains. North Carolina had somewhat recently developed an enormous pig factory farming industry, which was very hard hit by the hurricane-caused floods. As these storms sent local streams over their banks, untold quantities of pig manure and hundreds of thousands of drowned swine carcasses were distributed over the lowlands of eastern Carolina. In a matter of days, the dead swine began to rot. Groundwater was compromised for months afterward and homeowners who used wells—which were the majority of residents in these rural counties—had to make other arrangements for their water. The situation could have been much worse had the hurricanes struck earlier in the season and been followed by a few days of late summer heat.

Despite miraculous advances in medical technology, genetic typing, and immunology, the nations of the world are not much better prepared for a severe flu epidemic than they were for the 1918 outbreak. Epidemic influenza is extremely difficult to counteract. Flu vaccines developed in any given year are notoriously ineffective against new strains that come along the following year. It takes seven months or more to create, test, manufacture, and distribute a vaccine developed in direct response to a new virus, and by that time the disease can burn through global populations. If a pandemic broke out today, hospital facilities would be overwhelmed. Nurses and doctors would be infected along with the rest of the population.

Methods of factory farming in recent decades have included massive dosing of the animals with antibiotics; the predictable result has been the evolution of germs and bugs that are now resistant to drugs, in particular the bacteria responsible for food poisoning: *Salmonella*, *E. coli*, and *Campylobacter*. It takes years to develop, test, and gain approval for new antibiotic drugs. So while pharmaceutical companies are slowly developing potent new classes of antibiotics, resistance is developing at a rate faster than the drug companies can develop replacements. The overuse of antibiotics in livestock has been mirrored by the overuse of antibiotics in regular medicine. Within the last few years there has been an emergence of bacteria resistant to vancomycin—a last defense drug for some illnesses, including deadly blood infections and pneumonia caused by *Staphylococcus* bacteria.

Factory farming of animals has been behind the frightening and mystifying mad cow problem. The effort to economically hyperrationalize meat production on a gigantic scale led to the use of slaughterhouse waste in cattle feed as a protein booster. The material used included the brains and spinal cords of cattle, sheep, and pigs, turning livestock, in effect, into cannibals—and they are not even supposed to be carnivores. In England, where proportionately more sheep and lambs are raised than in the United States, sheep's brains and nerve tissue infected with the neurological disease scrapie made their way into cattle via commercial feeds, and the cattle began presenting horrifying symptoms—loss of motor control, raging fits, seizures, and ultimately death. Autopsy showed that the affected cows' brains were riddled with channels and holes, like sponges; hence the name of the disease: bovine spongiform encephalopathy (BSE). The disease first came to public attention in 1986. In the years since, 155 human beings, mostly in England, came down with an odd variant of a rare condition called Creutzfeldt-Jakob disease (CJD). Autopsy showed very similar spongy brain degeneration as had been found in BSE-infected cattle. CJD had previously been encountered as a medical curiosity in such exotic milieu as the more isolated parts of Indonesia, where there was a long tradition of eating the brains of enemies slain in tribal warfare. The CJD that showed up in Europe presented slightly different and terrifying symptoms. With this variant disease, called vCJD, patients showed not just dementia but also extraordinary behavioral problems—wild rages, violence, screaming. Unlike previously-known CJD, which almost always appeared in victims over sixty years old, the new variant showed up in younger adults. It was inexorably fatal. It was also believed to have an exceptionally long incubation period, longer even than the AIDS virus, somewhere between ten and twenty years. Because of the exceptionally dramatic course of the illness and the long incubation period, a very strong reaction set in once the public became informed of the problem. This led to the wholesale slaughter of British cattle and the collapse of the English beef industry. Years later, English beef is still regarded with suspicion in Europe.

The little that is known about the agent responsible for the disease ought to be worrying. The group of transmissible spongiform encephalopathies (TSEs), which includes mad cow, scrapie, and Creutzfeldt-Jakob, is believed to be caused by a rogue protein called a prion (pronounced PREE-on). They are not living organisms per se, not like bacteria, or even viruses (which are, arguably, mere bundles of RNA with a mission). Prions are just proteins that are "folded" differently. They appear to have the odd characteristic of entering the bodies of animals and getting other proteins to fold the way they do: from alpha helical structures relaxed into looser beta sheets. Eventually prions completely clog the infected brain cells. The cells misfire, work poorly, or don't work at all. The structure of the brain itself becomes visibly and grossly deformed. In 1996, British scientists reported that the prions found in vCJD patients were remarkably like those of the mad cows. Prions are notoriously difficult to destroy. They appear to survive incineration and the heat of medical autoclaves. They also appear to have the ability to persist in the soil of a given place indefinitely. It is believed that millions of Britons ate BSE-infected beef during the 1980s and 1990s. There has been a sustained rise in cases of vCJD into the early years of this century. The rate of increase is about 20 percent a year, though the total numbers have remained very low so far. Nobody knows how many people will ultimately be affected.

Dark Winter

In June 2001, the federal government carried out a war game at Andrews Air Force Base called Operation Dark Winter. The object of the exercise was to see how public officials and the public health system might react in the event of a terrorist biowarfare attack, though a severe influenza infection of the 1918 kind could have played out similarly. The biological agent of choice for the exercise was smallpox, a horrific viral disease of tremendous infectious power that the World Health Organization had heroically eradicated "in the wild" at the end of the twentieth century, but which still existed in lab specimens in Russia and the United States—and perhaps, anxious leaders feared, in the laboratories of rogue states such as Iran and North Korea.

Operation Dark Winter employed a cast of volunteers—including Governor Frank Keating of Oklahoma, former senator Sam Nunn, former presidential adviser David Gergen, former CIA director James Woolsey, and former FBI director William Sessions—to act out roles following a script in which a terrorist released smallpox in one eastern U.S. city. The result was sobering to an extreme. The public health system virtually collapsed. Hospitals degenerated into chaos. Smallpox spread to twenty-five states and overseas. The national stockpile of vaccines proved to be deeply inadequate. The exercise was called off after four days from the sheer exhaustion of the participants, while the fictional epidemic was still spreading.

This exercise took place three months before the September 11, 2001, attacks and the subsequent malicious release by parties still unknown of weapons-grade anthrax in post office facilities and a Senate office building, which confirmed the susceptibility of the United States to a biowarfare catastrophe. Given the two-week incubation period for a disease such as smallpox, there is nothing to stop a malicious "carrier" from going pretty much wherever he or she wishes—airports, sports stadiums—without anyone suspecting. International intelligence agencies have warned that known terrorist groups have attempted to purchase biowarfare agents, and the truth is that no one knows whether they have been successful. According to those participating in Dark Winter, the results of a bioterrorism attack on the United States would be massive civilian casualties, breakdown in essential institutions, violation of democratic processes, civil disorder, loss of confidence in government, and reduced strategic flexibility abroad.

While it is certainly an extreme notion, the possibility exists of governments or ruling elites strategically using "designer" diseases within their own borders to cull unwanted populations without incurring any political blame—though the social cost could be fantastic. The idea might seem outlandish, but then so were the Nazi extermination of the Jews, Stalin's murder of the kulaks, the depredations of Pol Pot in Cambodia, the wholesale butchering of Tutsis in Rwanda, the starvation of North Koreans under

Kim Jong Il, and many other atrocities carried out in the modern era. An elite in some nation with terrible population pressures might engineer a disease and a vaccine at the same time, inoculating fellow members of the elite while loosing an epidemic on its own citizens. The SARS virus was suspected of being a prototype in this category. Severe acute respiratory syndrome first appeared in Asia in February 2003, out of nowhere, related to the coronavirus typically associated with the common cold. It was much worse than the common cold: It eventually spread to many corners of the world and infected just over 8,000 people, of whom about one in seven died from the disease.

The germ theory, which emerged in the late nineteenth century, focused the world's attention on the specific agents responsible for particular diseases, but the social and ecological contexts are equally important, and these are now coming more prominently into play with world population well beyond the limits of the earth's "natural" carrying capacity and with climate change apparently in progress. Stress on ecological equilibrium, rapid changes in land use, penetration of formerly inaccessible habitats, and disturbed migration routes can lead to the appearance or diffusion of a disease. While we may be able to identify the microorganisms involved, we can be helpless in the face of it, and our behavior may still promote its spread. Lyme disease in America, *Borrelia burgdorferi*, is largely a disease of suburbanization. Suburbanization is a form of behavior. The spirochete responsible for Lyme disease exists in a relationship with field mice, deer, and a particular tick, *Ixodes dammini*. The populations of deer, mice, and the tick exploded over a fifty-year period in the edge habitats that suburbia presents. By a similar change in conditions in a much different context, malaria increased significantly in Malaysia when jungle was cleared for rubber plantations early in the twentieth century. Snail fever and malaria spiked markedly with the building of Egypt's Aswan Dam in the 1960s.

Certainly microbial traffic has increased enormously during the current period of hyperglobalization—to the degree that, for instance, the extreme exotic Ebola hemorrhagic fever, originating from an unknown disease reservoir in the African jungle, turned up simultaneously in 1989 at primate research facilities as far apart as Reston, Virginia; Alice, Texas; and the Philippines.

In the first stages of the Long Emergency, the conditions affecting the spread of new disease may be very confusing. For instance, climate change is certain to alter patterns of microbial traffic. But the disturbance of global oil markets as the permanent energy crisis begins is liable to interrupt global commerce and global travel. Fewer businessmen will fly from continent to continent. However, these same energy problems will surely reduce crop production, which would lead to reduced food aid to desperate populations in poor nations, which would then lead to compromised immune systems and the migration of poor, hungry, and probably unhealthy people—and by "migration" I do not mean the orderly entry of people through airport lines, but rather the uncontrolled rush of desperate mobs, tribes, and whole ethnic groups from failing habitats into lands already occupied because they can better support human life. This is an obvious recipe for conflict and woe. Where the refugee camps set up, disease will surely follow.

The Social and Economic Consequences of Disease

The attrition of global populations by disease may be unavoidable. Some readers may regard it as the inevitable revenge of nature against the hubris of a human species arrogantly exceeding the carrying capacity of its habitat. Some may regard it as a moral victory against wickedness. Some may view it in the therapeutic mode as a positive development for the health of the planet. Many self-conscious "humanists" have militated for the goal of reducing population growth—though most of them would have probably preferred widespread birth control to a die-off. But that kind of thinking might have been just another product of the narcotic comfort of cheap oil, as merely stabilizing the earth's population at current levels (or even 1968 levels) would arguably still have left humanity beyond the earth's carrying capacity. Apart from these issues of attitude and ethics, however, a major decline in world population, or change in demographic profiles, is apt to have profound and strange repercussions on everyday life.

The bubonic plague or Black Death of the period 1347–51 killed off a third of Europe's population. Medieval society never recovered. The plague killed entire families and destroyed at least 1,000 villages. It tended to kill proportionately more of the lower classes who lived in greater squalor and did not enjoy the ability to remove or quarantine themselves from plague hot spots. But at the same time, the plague erased the chief advantage of belonging to the upper class in the first place—access to ultra-cheap labor. The plague drastically altered economic and social relations. Civil disorder followed the plague in its march up through Italy, France, England, Germany, and Scandinavia. Banditry and lawlessness were commonplace wherever the plague burned. Authority was unable to function decisively. Public health measures were ineffective because the source of the plague was not understood—strewing herbs and murdering Jews did nothing to improve the situation. Faith in a merciful Christian god eroded, and the collective psychology of the survivors swooned into morbidity and depression. The art of the period, with its cavorting skeletons and corpses, reflects this.

The plague of 1347–51 had been preceded by a general European famine that began around 1315, when the climate of Western Europe wobbled and produced a series of late springtimes and cool, wet summers. The growing season was substantially shortened. Seed varieties that could withstand different conditions were not available. Livestock suffered and ergot fungus attacked the reduced grain harvests. Wine production was severely affected, especially in southern England, where it never recovered. The Viking settlements of Greenland withered away.

Western Europe had been a forgotten backwater of the known world after the fall of Rome and the consequent shift of wealth and power to distant Constantinople. A brief climate cooling had accompanied the fall of Rome and its aftermath. Europe had endured centuries of darkness, cultural amnesia, and squalor. The climate then underwent a general warming from about A.D. 900 to 1300. Life in Europe improved. Under the mode of social organization generally labeled feudalism, European populations increased along with the food supply. Much of the surplus wealth that the feudal kings and lords of Europe managed to acquire was spent in the ongoing project of the Crusades, an attempt to defend Christianity by pushing back the conquests of a militant Islam that had subsumed the old Christianized people of the Middle East, then moved aggressively through North Africa up into Spain and France, and also into Christian Asia Minor by way of the Seljuk Turks. For three centuries the armies raised for the Crusades also had been an outlet for a European peasantry multiplying under favorable conditions, and thus a brake on population growth.

By 1291 the Crusades were over, ending in a stalemate, with the Muslims shut out of northern Europe and the Europeans chased out of the Holy Land. Over the next peaceful quarter-century Europe's population reached a critical level about equal to the solar carrying capacity of the region. That is, the people could raise enough food to feed themselves and no more. Some towns were beginning to suffer from scarcity of firewood. The tweak of climate change beginning in 1315 lowered the carrying capacity of Europe instantly. Grain production suffered markedly for three years running and a general famine commenced. Even when "normal" weather patterns returned after 1318, there was a scarcity of seed grain to resume full food production and the famine lingered. The mortality rate was high and all classes eventually suffered. Ten to 15 percent of the population died, most from disease induced by weakened immunity. The famine certainly provided a vivid and tangible sense of limits for the number of people the region could support, a warning from the earth to its inhabitants that was, of course, interpreted as a punishment visited by God for man's wickedness.

By 1325, agriculture in Europe had recovered, and for the next twenty years the population resumed rising back to the solar carrying capacity of the region. Before long, the military effort that had been put into the Crusades for so long was now directed into the first skirmishes of the Hundred Years' War, a contest between England and France over the control of French territory. This was explicitly a struggle for extra carrying capacity and resources.

The chief beneficiaries of the Crusades had been Venice, Genoa, and the other Mediterranean ports. And it was as a result of the trade these port cities generated with far-flung corners of the world that the bubonic plague stole into Europe from somewhere in Asia in 1347. What made the plague so terrible was not just the sordidness of the disease itself, or even the shockingly high mortality rate, but the fact that once established it recurred in the same region or city for several years running. So many people died that there were labor shortages all over Europe. By the end of the 1300s peasant revolts broke out in England, France, Belgium, and Italy.

Feudalism, based on a surplus of agricultural peasants tied to a particular place, unraveled. The general notion of wealth and status began shifting from land to money. Though urban areas suffered grievously during the plague years, in the aftermath of the epidemic displaced peasants and rural artisans gravitated into depopulated towns and cities, found opportunities there, and began to take part in the civic relations that would lead to a new commercial society we now identify with the Renaissance.

A contemplation of these circumstances that occurred seven hundred years ago gives us an idea of what to expect in the Long Emergency. One big difference is that now we can see it coming. However, we in America flatter ourselves to think that we are above this kind of general catastrophe—because our technologic prowess during the cheap-oil fiesta was so marvelous that all future problems are (supposedly) guaranteed to be solved by similar applications of ingenuity. This was certainly the consensus among the scientists, computer geniuses, and biotech millionaires I rubbed elbows with this year at get-togethers such as the Pop Tech conference. They were uniformly uninterested in the issues of the global oil peak and natural gas depletion and utterly convinced that the industrial societies would be rescued by hydrogen, wind power, and solar electricity, all to be figured out by their cohort techno-geniuses in due time. If there is anything we have been stupendously bad at in the preceding century of wonders, it is recognizing the diminishing returns of our technologic prowess. Some of our greatest achievements, such as industrialized farming and the interstate highway system, have produced dreadful diminishing returns (e.g., national epidemics of obesity and diabetes and the fiasco of suburban sprawl). This persistent failure or weakness pretty much negates the value of our ability to see what's coming. If anything in the turbulence of the Long Emergency, rather than technologic progress, we are more likely to see a lot of technologic regress—the loss of information, ability, and confidence.

The coming crisis over oil and natural gas will be bad enough where American food production is concerned, and would in itself be enough to pose a grave threat to society. The additional threat of climate change can only intensify the difficulty of reforming American agriculture back in the direction of smaller-scaled, local farms free of oil and gas dependence. All other sectors of the American economy (and its global context) will be deeply affected by the oil predicament and climate change. The car-dependent infrastructures of suburbia will become progressively unusable, and with them many economic activities. Trade networks based on the assumptions of permanent globalism will cease to operate. Electric service may not be reliable in the way that we're accustomed to it. Occupational niches will vanish on a massive basis, and with them, livelihoods. New classes of economic and social losers will be created. Their anger will generate political trouble within the developed nations, especially in the United States, as I will spell out later. Nations, regions, factions, and tribes will fight over declining resources, and will unfortunately expend ever more precious human and economic capital in the process. Governments may become increasingly impotent, at the national level in particular, and the rule of law may be suspended in some places, as it was during the crisis of the Black Death.

Many individual immune systems will be compromised by the hardships of the Long Emergency and disease will seize the opportunities presented, as it always has. AIDS ought to be especially worrisome, because even when people have lost everything, they still have sex. That may be all many people will have, and it will get them in a lot of trouble. Besides, as already suggested, the resourceful HIV bug may find an even more efficient means of transmission through countless random acts of mutation. Millions of human beings are going to die. I have no idea what the population of survivors might be. The attrition is apt to continue for much longer than the Black Death raged in the Europe of the fourteenth century, because under the regime of cheap oil the carrying capacity of our earthly habitats was exceeded by orders of magnitude, and we have farther to go to return to the solar carrying capacity of our home places. Some home places, such as the deserts of Arabia and the American West, will support only minuscule numbers of people without the benefits of fossil fuels. Of course there will be no compensations for the loss of those nonrenewable resources. Also, because of the probable human contribution to global warming, this climate change might well be much more severe and longer-lasting than the blip of the early 1300s, or even the Little Ice Age of the seventeenth and eighteenth centuries.

As hunger and hardship increase, the world may see more than one wave of more than one disease. If and when an influenza pandemic emerges, for instance, many AIDS sufferers will succumb, but people infected with the AIDS precursor, HIV, will still survive influenza and AIDS will march on. India, for example, was among the hardest-hit nations in the 1918 flu pandemic. Today it has among the highest rates of AIDS infection. The age-old human enemies, tuberculosis, malaria, cholera, streptococcus, and other members of the familiar gang, will be on hand with new immunity to the old techno-tricks of the twentieth century. Even after these diseases may have spent themselves for a while, climate change will still be with us. Nobody really knows where that is taking us, though we do know that the human race has endured more than one ice age in the past.

The current urban population of the world, 3.2 billion, is greater than the entire population of the world in 1960. Seventy-eight percent of the urban dwellers in the so-called developing world live in slums. From the West African littoral to the mountain sides of the Andes to the banks of the Nile, the Ganges, the Mekong, and the Irrawaddy, new gigantic slums spread like immense laboratory growth media, waiting to host epidemic disease cultures. Lagos, Nigeria, for example, grew from a city of 300,000 in 1950 to over 10 million today. But Lagos, writes Mike Davis, "is simply the biggest node in the shanty-town corridor of 70 million people that stretches from Abidjan to Ibadan: probably the biggest continuous footprint of urban poverty on earth."[10] Most of the world's new, exploding slums have only the most rudimentary sanitary arrangements, open sewers running along the corridorlike "streets." In the slums of Bombay, there is an estimated one toilet per five hundred inhabitants. Currently two million children die every year from waste-contaminated water in the world's slums. The enormity of this urban disaster is poorly comprehended in advanced nations like the United States, where the drinking water is still safe and even the poor have flush toilets connected to real sewers. But the slums of the world will probably be the breeding ground of the next pandemic, and chances are, once it is under way, the wealthy nations will not be spared.

It ought to be pretty obvious that the social systems, subsystems, and institutions necessary to run advanced societies would be weakened, perhaps beyond repair, by the multiple calamities of the Long Emergency.

10. Mike Davis, *Dead Cities, a Natural History*, Los Angeles: New Press, 2002.

These were the implications of Operation Dark Winter, and they are the most visible symptoms of life in the disintegrating postcolonial pseudo nations of AIDS-stricken Africa. A variation on this theme has played out in post-Soviet Russia and its former vassal states, now spiraling into destitution. The sum of all these subsystems and institutions that support a complex society add up to an economy, or a political-economy, to use a somewhat antique term that better captures the connection between the activity of a social system and its governance. The nature of the political-economy in which we find ourselves now, on the brink of the Long Emergency, and the kind that we will find ourselves in during the Long Emergency, and how people may negotiate that difficult transition, will be the subject of the last two chapters of this book.

Six

Running on Fumes:

The Hallucinated Economy

The most significant characteristic of modern civilization is the sacrifice of the future for the present, and all the power of science has been prostituted to this purpose.

—William James

The entropic mess that our economy has become is the final blowoff of late oil-based industrialism. The destructive practices known as "free-market globalism" were engendered by our run-up to and arrival at the world oil production peak. It was the logical climax of the oil "story." It required the breakdown of all previous constraints—logistical, political, moral, cultural—to maximize the present at the expense of the future, and to do so for the benefit of a very few at the expense of the many. In America, free-market globalism became the reigning orthodoxy of both political parties, challenged only by cranks wearing nose-rings at the very margins of society. The moment that the world recognizes the passing of the oil production peak as a reality, globalism will be dead both in theory and practice.

During the years of its brief reign, free-market globalism was regarded as a permanent institution by a broad consensus of leaders from the most august Harvard economists to the most vulgar corporate buccaneers. The news media and their left-right punditry all bought it, too. The idea was that humanity had arrived at an advanced level of sociopolitical evolution, a new economy that would eventually deliver heaven on earth, where everyone everywhere would be rich. The key word was "eventually."

Globalism pretended to promise the same nirvana as communism had failed to deliver in its time, and came into full flower just as communism lost its legitimacy. Globalism also had the same tendency to impoverish and enslave huge populations while enriching the elite who managed its operations. The American people were sold on it, even while it destroyed their towns, their landscapes, and their vocations. What a shock, then, to find out that the so-called global economy was just a set of transient economic relations made possible by two historically peculiar circumstances: twenty-odd years of relative international peace and reliable supplies of cheap oil.

Who Needs the Future?

Globalism was primarily a way of privatizing the profits of business activity while socializing the costs. This was achieved by discreetly discounting the future for the sake of short-term benefits. The process also depended on the substitution of corporate monocultures and virtualities for complex social ecosystems wherever possible, for instance, Wal-Marts and theme parks for towns. Globalism was operated by oligarchical corporations on the gigantic scale, made possible by cheap oil. By "oligarchical" I mean that power was vested in small numbers of people running large organizations who were not accountable for their actions to many of the people who were subject to those actions. By "corporation," I mean a group enterprise given the legal status of a "person," with "rights," but in fact devoid of any human qualities of ethics, humility, mercy, duty, or loyalty that would constrain those rights. As Wendell Berry put it, "a corporation, essentially, is a pile of money to which a number of persons have sold their moral allegiance. . . . It can experience no personal hope or remorse. No change of heart. It cannot humble itself. It goes about its business as if it were immortal, with the single purpose of becoming a bigger pile of money."[1] The corporate oligarchs of, say, Wal-Mart, Archer Daniels Midland, and the Disney Corporation were not necessarily evil people, but it

1. Wendell Berry, "The Idea of a Local Economy," Orion Online (*www.orion. com*), 2002.

was in the nature of their actions that a great deal of harm came to localities and local people in them. Under the banner of free-market globalism, the chief side effect of oligarchical corporatism making its money piles bigger was the systematic destruction of local economies and therefore local communities. Thus, the richest nation in the world in the early twenty-first century had become an amazing panorama of ruined towns and cities with broken institutions and demoralized populations—surrounded by Wal-Marts and Target stores.

The free-market part of the equation referred to the putative benefit of unrestrained economic competition between individuals, and because corporations enjoyed the legal status of persons, they were assumed to be on an equal footing with other persons in a given locality. Thus, Wal-Mart was considered the theoretical equal of Bob the appliance store owner, and if Bob happened to lose in the retail competition because he couldn't order 50,000 coffee-makers at a crack from a factory 12,000 miles away in Hangzhou, and receive a deep discount for being such an important customer, well, it wasn't as though he hadn't been given the chance.

The free market also referred to an extreme version of the old idea of comparative advantage, which had meant originally that every locality has something special it is good at producing, or some raw material in ready supply, and that a larger macro economy is made up of such specialist trading partners. Under globalism, this was modified to mean that for the sake of "efficiency" such trading partners ought to forget everything else and pump out as much of their specialty as possible (using the money received to buy goods and services from other specialists). There were a number of problems with this simplistic idea. One was that cheap oil subsidized the whole system, and the system would have been impossible without it.

Cheap oil had allowed populations to explode in precisely those parts of the world that had had, for millennia, a high infant mortality rate and modest life expectancy. Cheap oil was behind the "green revolution" that increased the food supply in the nonindustrial world. Oil was also behind many of the medicines and preventives that had neutralized tropical diseases. Now, suddenly, most of those children actually survived, grew up, and produced more children who survived and grew up, and over the course of the twentieth century, the global populations hurtled into extreme numerical overshoot. Populations were, in effect, eating oil, notably in food exports from the United States, where agribusiness had completely taken over from agriculture. Local farmers in Africa, Asia, or South America couldn't compete with corporate Archer Daniels Midland's oil-and-gas-based grain crops and U.S. government subsidies. There was no point in even bringing their hardscrabble crops to market when sacks of cheap American wheat sat on the docks of Pusan or Colombo. Farmers in those places felt that they had no choice but to migrate to the city and find some other way to get by. The only comparative advantage that these people possessed was their willingness to work for next to nothing. Cheap oil and free-market globalism turned comparative advantage into a new kind of feudalism, with the corporations as the lords and the overabundant locals as the serfs. And then, when the comparative advantage of cheap labor ($5 a day) of one place, such as Mexico, was superseded by the cheaper labor (99 cents a day) of another place, such as Sri Lanka, the corporations just moved their operations.

The idea of comparative advantage works when there is a complex local economy intact in the background of each trading partner's specialized item of production, with a variety of social roles and occupational niches to support the long-term project of community. But a locality geared to doing only one thing for export is ultimately a slave system based on the extractive economics of mining. In the extreme version of comparative advantage, under the regime of hyper-turbo late-oil-age industrialism, with its ultracheap transport and instant communications that defeated any advantages of geography, the only comparative advantages left were cheap labor and free capital. One group had all the cheap labor and another group had all the capital, and for a while one group made all the things that the other group "consumed." Thus, comparative advantage became, for a time, a con game strictly for the benefit of large corporations, which ended up enjoying all the advantages while the localities sucked up the costs.

The corporations benefiting from this regime often had no physical home of their own, even in their country of origin—and not a few American corporations had moved their official address to Caribbean pseudonations, where the banking and tax laws were more agreeable. The corporations had no allegiance to any particular place or the people of that place, so the destruction they wreaked was as manifest in the ravaged towns of Ohio and upstate New York as in the environmental degradation of China. America was hardly immune to the consequences of free-market globalism. In effect, the American heartland was overtaken by a new kind of corporate colonialism, emanating from our own culture, but no less destructive than the imposition of foreign rule.

Americans failed to recognize the essential fraudulence of the idea that this destruction was "creative" and would lead to a higher good—in other words, that the end justified the means, even as they watched their towns die around them. Corporations such as Wal-Mart and its imitators used their wealth and muscle to set up "superstores" on the cheap land frontier outside small towns and put every other retail merchant out of business, often destroying most of the town's middle class. They also, incidentally, destroyed the local capacity to produce goods. And the American public went along with it for the greater good of paying a few dollars less for a hair dryer. Bargain shopping justified the extermination of the middle class and all its relations with the locality. The American people were gulled into the fantasy that every day of the year would be like Christmas, Wal-Mart style. The public enjoyed this bonanza of supercheap manufactured goods without reckoning any of the collateral costs, which were astronomical.

The local merchants who were put out of business had been the caretakers of the town. They often owned at least two buildings in town—their homes and the buildings in which they did business—and they generally took good care of them. The physical decrepitude that is now the most visible characteristic of American towns is the direct result of extirpating that class of local people. These individuals also were generally the caretakers of the town's institutions. They sat on the library boards, the school and hospital boards, the planning board. They ran the local charities. They were invested in the history of a place and their living actions had to honor the memory of their forebears and the prospects of generations to come after. Every virtue that grew out of these local relations of person and place was traduced by the big-box national retail corporations, and the American public was absolutely complicit in the hosing that it got.

This raises an interesting question: Is one led to a determinist view that this outcome was an inevitable result of circumstances? Did Americans sell out their towns, their neighbors, the memory of their ancestors, and the future of their grandchildren because they were helplessly in thrall to the blandishments of a cheap-oil economy? I honestly don't know, though I tend to view the outcome as the result of many collective bad choices made by the public and its leaders. But were those choices inescapable? Certainly the process was insidious and played out over several generations.

There is a kind of narrative arc to a story like the industrial revolution. It had a beginning that is fairly easy to establish, say, from Newcomen's first steam engine in 1725, deployed to pump water out of the British coal mines. It had a middle, which I put around 1900 with the factory system fully established, cities at their peak of development, and the conversion from coal to petroleum under way. We are reluctant to identify the climax or the beginning of the end because we are afraid that we will suffer in it, and it is very hard to imagine a world without technological amenities, or fewer of them, or the process in which they become lost to us, and our comfort and safety perhaps with them. In any case, I'd propose that the industrial "story" climaxed in America during the 1970s. The climax was coincident with our passing of the American oil peak. And what we have been experiencing with the so-called free-market global economy and all the disruptive damage around us is a manifestation of our slow and painful arrival at the end of the story. This final phase has taken about thirty years so far, and will probably be complete within this decade. And it will certainly be coincident with the passing of the global oil peak. The economy of the past three decades has been increasingly freakish and bears some examination.

The High-Entropy Economy

The industrial revolution had already begun when the United States formed in the late 1700s, and even before Americans exploited coal, oil, and gas, we had other resources to plunder, notably land. The American economy has historically been about moving incrementally away from natural patterns of living off solar energy to artificial patterns of living subsidized by cheap fossil fuel. Through the tumultuous twentieth century to now, the American economy has moved insidiously from the sustainable to the terminal, from the solar flow to the nonrenewable stock, from the authentic to the virtual, and from the actual to the abstract. All these transactions embody a particular loss of one kind or another, but they add up to a general loss of the potential to continue the way we have been living. This loss can be described as entropy.

The First Law of Thermodynamics says that energy cannot be either destroyed or created, only changed. Entropy, the Second Law of Thermodynamics, says that the change of state in any given amount of energy flows in one direction, from being concentrated in one place to becoming diffused or dispersed and spread out; from being ordered to being disordered. A hot cup of coffee cools off sooner or later. Its heat is diffused until the temperature of the coffee stabilizes to equilibrium with the air around it. It never gets spontaneously hotter. A tire goes flat, it never spontaneously reinflates. Windup clocks wind down, they don't wind up. Time goes in one direction. Entropy explains why logs burn, why iron rusts, why tornadoes happen, and why animals die.

The reason that everything in the real world does not fall apart at once is that the flow of entropy faces obstructions or constraints. The more complex the system, the more constraints. A given system will automatically select the paths or drains that get the system to a final state—exhaust its potential—at the fastest possible rate given the constraints. Simple, ordered flows drain entropy at a faster rate than complexly disordered flows. Hence, the creation of ever more efficient ordered flows in American society, the removal of constraints, has accelerated the winding down of American potential, which is exactly why a Wal-Mart economy will bring us to grief more rapidly than a national agglomeration of diverse independent small-town economies. Efficiency is the straightest path to hell.

Inefficient economies are much more complex than efficient ones. Complexity itself can be deceiving. Biogenic complexity constrains entropy flows with checks and balances. What we take to be man-made artificial complexity (technology) is, paradoxically, a simplification process that increases flows by editing away inefficiencies. The ecology of a prairie will keep the soil active and healthy indefinitely, while the ecology of a fossil-fuel-subsidized cornfield will leach the soil of useful nutrients and physically erode it in less than a human lifetime. The ecology of a pond, with its diverse hierarchies of life and multitude of biological niches and food chains, is much more complex than the Crown Point, New York, trout hatchery with its monoculture of fish, its inputs of manufactured fish food, and its staff of attendants cleaning waste out of the cement hatchery impoundments. The natural pond also has more chance of continuing indefinitely into the future. The built-in constraints of inefficient biogenic economies reduce the flow of potential, often to the point where systems based on inefficient economies last for geologic epochs, not just a few decades in the case of a fish hatchery. Everything that we identify with nature takes the form of inefficient systems. Biogenic or living systems are self-stabilizing. They are self-buffered. Small differences are dampened out. Entropy is stalled within them. They exhibit negative feedback tending toward long-term stability. Call this condition "negative entropy." Everything we identify with the man-made substitutes for natural bioeconomies, that is, technologies, tends toward positive feedback, which is self-amplifying, self-reinforcing, and destabilizing, featuring the removal of constraints to entropy flows and leading to the certain eventual destruction of that system. Call this condition "positive entropy."

Positive feedback sounds like a good thing—like a reward for doing good deeds—but in terms of this discussion it has a different meaning. Positive feedback means that a given activity in a system is reinforced by the activity itself—nicely illustrated in Brian Czech's metaphorical book title, *Shoveling Coal on a Runaway Train.*[2] The most obvious self-reinforcing positive feedback mechanism in conventional economics is "growth." Growth has been the chief defining characteristic of our industrial economy. In the simpleminded commentary of the news business, growth is good (prosperity) and nongrowth is bad (depression). The more growth, the more business, the more prosperity, and vice versa. Econometrically, growth must be from 3 to 7 percent a year, or society is in trouble. The boosters of this particular notion of growth—namely, most mainstream economists—do not recognize any limits to growth projected into the future. They'd prefer not to think about it because the conclusion is obvious: There have to be limits. If we project "housing starts"

2. Brian Czech, *Shoveling Coal on a Runaway Train*, Berkeley: University of California Press, 2000. This is an excellent disquisition of the differences between "ecological" and "neoclassical" economics.

ninety-nine years forward at current rates, there wouldn't be a single buildable quarter-acre lot left in the world. Not a few economists would rationalize this outcome by declaring that ninety-nine years from now we will have colonies on the moon or Mars or under the Sea of Cortez. Or that technology coupled with human ingenuity will solve the problem some other way, perhaps by genetically reengineering human beings to be one inch tall, or booting all our consciousnesses into computer servers where unlimited numbers of virtual people could dwell in unlimited virtual environments of endless cyberspace.

More likely, we will remain confined to the planet Earth. Economic growth that has appeared normative and desirable during the story of industrialism is already becoming pathogenic in an economy showing more and more signs of positive feedback and accelerating positive entropy manifesting as damage to the biosphere. High entropy becomes particularly problematic in an economy utterly dependent on a few very special commodities: oil, natural gas, coal, and uranium. It becomes especially relevant when the limits to those commodities become tangible, as is now the case as we approach the global oil production peak and the actual depletion (thirty years past peak) of the North American natural gas endowment. But the collective imagination of the public cannot process the notion of a nongrowth economy, even though the limits to growth are visible all around us in everything from the paved-over suburban landscapes, to the steeply rising gas prices, to played out aquifers, to the death of the Atlantic cod fishery. We are not capable of conceiving another economic way. We are hostages to our own system.

The picture is further clouded by the notion of substitutability, a doctrine based on the observation that the sensitive device we call the market seems to call forth new resources as old resources become problematic (usually expressed in terms of higher prices). Hence, when trees grew scarce in England during the Little Ice Age (1560–1850), people there began to use more coal to keep warm, which caused people to dig deeper for it, which called forth the innovation of the steam engine to drain water from the mines so the miners wouldn't drown. However, an interesting positive feedback loop was set in motion. The invention of the steam engine (a magical product of human ingenuity) provoked the invention of other new machines, and then of factories with machines, which prompted the need for better indoor lighting, which stimulated the use of petroleum, which produced brighter light than candles (and was much easier to get than sperm whales), which provoked the development of the oil industry, whose oil was found to work even better in engines than coal did, which led to the massive exploitation of a one-time endowment of concentrated, stored solar energy, which we have directed through pipes of various kinds in an immense flow of entropy, which has resulted in fantastic environmental degradation and human habitat overshoot beyond carrying capacity. It is assumed now that human beings, prompted by the market, will employ ingenuity to discover a substitute for oil and gas, once the price starts to ramp up beyond the "affordable" range. This assumption is apt to prove fallacious because it ignores the fact that the earth is a closed system, while the laws of thermodynamics state that energy can't be created out of nothing, only changed from low entropy to high entropy, and that we have already changed the half of our oil endowment that was easiest to get into dispersed carbon dioxide, which is now ratcheting up global warming and climate change, which might well put the industrial adventure out of business before human ingenuity can come up with a substitute for oil. The solar energy stored for millions of years in oil will now be expressed in higher temperatures, more severe storms, rising sea levels, and harsher conditions for the human species, which, despite its exosomatic technological achievements, remains a part of nature and subject to its laws.

Finance, or Abstracted Economy

Money is a wonderful thing. It started out in human history as hard currency, generally gold or silver. These are commodities that are deemed to have intrinsic value but also act as a means of abstractly representing wealth accumulated out of other real commodities. Relatively little hard currency ever circulated freely in the preindustrial world. In that world, most wealth was actual, existing in the form of land, palaces, fleets of boats, bolts of cloth, barrels of grain, standing timber, herds of cattle, and so forth. These were generally things that could be traded, and exchanges of these items were often facilitated through the medium of gold or silver, hence the term "medium of exchange." Hard currency could also be acquired by theft or plunder, though that did not necessarily affect its value. Note that the value of hard currency is transcultural. Gold and silver had high value to the Europeans, the Chinese of the Sung dynasty, the Inca, the Aztecs, the ancient Egyptians, and the California Forty-niners.

The industrial experiment took the idea of currency (money) to the next level of abstraction—as hard currency can represent actual goods, so paper currency can represent hard currency and actual goods. As trade increased and took place over ever-greater distances, paper promises to pay hard currency began to steadily take the place of the hard stuff itself, which was cumbersome, hard to lug around in large quantities, and subject to theft in transit. So to streamline these trades, all kinds of certificates were used as equivalents to hard currency: individual ICUs, bills of lading, letters of credit from rich people, promissory notes issued by guilds. In time, the use of paper certificates became more and more normative and conventionalized. Protocols of exchange were established. Institutions were created to process them. This process of managing monetary affairs—of wealth abstracted in paper—was called *finance*.

The joint stock company arose largely as a response to the European discovery of the New World, and the huge expansion of trade it provoked, as a way of funding real-life speculative ventures beyond the scope of an individual's resources. The Pilgrim settlement of Massachusetts was such a venture, issuing shares to members in order to procure the necessary equipment, sailing ships, guns, shovels, axes, blankets. The division of shares, as well as a means for raising cash, was also partly a way of assigning a money value, or monetizing, the risks associated with making the harsh sea journey to a New World fraught with hazards, and in doing the hard work of settlement in order to gain productive land and thus future wealth. "Shares" in such a company were not easily transferable from one person to another, as their value was linked intrinsically to the particular individual who participated in a joint venture. But they did confer the benefit of allowing individuals to be part of a venture they could not possibly have carried out on their own.

Another type of incorporated venture was established by royal charter for organizing the extraction and trade of resources in faraway lands. The prototypes were the British East India Company and the Hudson's

Bay Company. Unlike settlement ventures, their shares were fungible, that is, convertible into currency and transferable from any one person to another at a negotiable (market) price. The next step was a corporate entity that government could use to combine enterprise in trade and resource extraction with the abstract procedures of capital formation to produce and manipulate paper currency. The exemplar of this was the Company of the West, the Enron of its day.

The strange history of the Company of the West begins in France roughly a hundred years after the Pilgrims established their Massachusetts Bay Colony. In 1715 King Louis XIV was dead and the French royal treasury was in a shambles after the Sun King's splendiferous and expensively long reign. The heir to the throne was his five-year-old great-grandson, crowned Louis XV. Obviously incapable of ruling, the tot's official duties were shifted to a regent, his great-uncle, the Duke of Orleans. Enter the curious, charismatic Scotsman John Law (1671–1729), a sometime banker, gambler, swindler, rake, and adventurer forced to flee the British Isles after killing an opponent in a duel.

Law escaped to France and managed to make his way into the court at Versailles. There he dazzled the Duke of Orleans with ideas for the repair of the nation's tattered finances—the government was hopelessly in debt and taxes were dangerously high. Law was liberal with advice, through which he saw many opportunities for personal advantage. Among his talents and accomplishments, Law had conceived some innovative theories about money and the uses to which it might be put. He argued that money was one and the same thing as credit, and vice versa, and that the amount of credit extant in a given society was determined solely by the *perceived* needs of trade. Credit could be dissociated from bags of gold locked in the basement of a royal treasury. It could be created sui generis, more or less by agreement among people that it was needed for a particular purpose, perhaps only an imagined purpose. And institutions could be constructed to facilitate the process. Law essentially made clear the distinction between a passive treasury, where money just accumulated, and an active bank, where money was *created.* Law's advice yielded the establishment of a state-chartered Banque Générale with the power to issue paper notes in 1716. Law benefited handsomely as presiding executive. The following year, Law also took control of a languishing venture called the

Mississippi Company, organized some years earlier to develop the then-French colony of Louisiana in North America, but left adrift after the Sun King's demise.

The Louisiana territory was much bigger than France proper and was believed to contain untold riches in gold, furs, timber, and productive land. The potential for economic exploitation was stupendous, everyone agreed. To date, though, the trade in these commodities had been poorly organized, and the French government was enjoying little benefit from the wealth extracted. Law modeled his reorganization of the company along the lines of the British East India Company, and induced the French government to grant the Mississippi Company the exclusive charter to control all economic operations in this immense region. Law sold shares in the company for cash and for bonds issued by the national treasury. Law's Banque Générale looked to be such a success that the French government made it the national bank and renamed it the Banque Royale. Law remained its director, enjoying substantial emoluments. Through this position he gained control of other companies trading in Africa, China, and the East Indies—in effect, France's trade with the rest of the world—and melded them into the Mississippi Company, the world's first modern super-conglomerate.

In 1719, Law devised a scheme in which the anticipated wealth generated by the Mississippi Company subsumed the entire French national debt, and he launched a plan whereby portions of the debt would be exchanged for shares in the Mississippi Company. In 1720, Law was appointed controller general of the national department of finance. He had the power to mint coins and collect taxes, in addition to controlling the nation's trade and its colonial ventures. The value of Mississippi Company shares soared. New World bonanzas had sufficient reality to persuade the skeptical. The world had already seen Spain seize the treasure houses of Montezuma and the Inca. The English had thriving colonies from Maine to Virginia. Nobody really knew what Louisiana held, but at the very worst it meant a lot of good cheap land for sale. The nascent middle class poured their limited savings into the company and other foreign nationals rushed into buying shares as well. A mania in Mississippi Company stock was under way. Value of a share rose from its initial offering of 500 livres to 18,000 in a year.

Law got into trouble when his bank, now the national bank, issued more and more paper money to fund purchases of Mississippi Company stock. It was all consistent with his general theory of money and credit: The market (the Mississippi Company) demanded investment so money (banknotes, bonds) were "created" to service that need—all based on the reasonable expectation of the company producing future huge riches. Law's ideas were rational. But it was not generally understood then, in the childhood of modern finance, that where theory might be perfectly rational, financial markets are generally highly irrational, subject to two powerful human emotions, fear and greed. In other words, markets do not function with perfect efficiency, either in terms of assigning monetary value to future prospects, or in terms of producing durable agreements as to that value. *Tant pis.*

Law's financial edifice began unraveling quickly in 1720, as soon as he took office as controller general. At the height of the Mississippi Company stock's value, investors started a run of profit-taking, selling their shares for gold. Demand to sell became so brisk that Law, in his capacity as head of the national bank, limited gold withdrawals from the treasury to a fraction of share value to halt the selloff. This maneuver only converted greed into fear. Shares could be exchanged for banknotes, but the population was wary to accept them as real money. Suddenly, it mattered again that banknotes needed to be backed by hard reserves. The bubble burst in May 1720 when a run on the Banque Royale forced the government to acknowledge that the amount of precious metal in the vaults was not quite equal to half the total amount of paper currency in circulation. Panic set in and the share price dropped as dramatically as it had risen, while banknotes lost their value. The cycle was implacable. By November of that year, the shares were worthless and Law was forced to flee the country.

The financial ruin of the French middle class who had invested so broadly in Mississippi Company shares set off an extreme and persistent disaffection between them and the nation's elite that eventually brought down the French monarchy in 1789. Actual settlement of Louisiana never really got off the ground and the region was swapped around between England, Spain, and France for the remainder of the century. Eventually, Napoleon regained control of the territory (after promising Spain substantial control of Italy), but turned around and sold it to the United States, whose citizens were now settling the edges of Louisiana by default anyway. The idea of paper money stood discredited in France and the country lagged in readopting banknotes (or modern methods of finance generally) just as the industrial revolution began in earnest in England. This failure to establish methods for deploying capital also helped both provoke and protract the French revolution. The word "banque" itself remained in such disrepute that to this day many French banks such as the Credit Lyonnaise still don't even use it in their names.

A parallel swindle in England at the very same time was organized around the South Sea Company, which sold shares on a similar pretext as Law's Mississippi Company—a monopoly on New World trade. The South Sea bubble was a more overt pyramid scheme than the Mississippi Company and the British government was merely one duped party among many than a cosponsor of the fraud, as France was in the case of Law's shenanigans. When the South Sea Company finally went up in a vapor, it ruined many families and injured the reputations of eminent men, including Sir Isaac Newton, but it did not destroy the credibility of British finance per se, much to England's later advantage in jump-starting the industrial age.

Though he was a figure of bizarre character and destiny, Law's financial ideas were not altogether dishonorable and certainly were touched by genius. The Louisiana territory was fantastically rich, if not strictly in gold, but its development under the best circumstances would have taken generations, and the Mississippi Company "bubble" lasted barely a couple of years. Meanwhile, Law's notions about financial credit issued by central banks backed by a consensus about the value of a productive society rather than hoarded bags of gold, and the deployment of that credit to create new wealth, were a precursor both of industrial capitalism and the eventual melding of industrial capitalism and activist central government as epitomized in the ideas of John Maynard Keynes, which became the orthodoxy of the twentieth century. After the Mississippi bubble fiasco, Law himself resumed his wanderings around Europe as a gambler and died penniless in Venice eight years after the collapse of his ventures in France.

First Heyday of the Corporation

The elevation of abstract finance as a valid realm above the "real" world of hard assets and actual commodities gained legitimacy as the industrial revolution advanced. As commodities and finished goods multiplied, the amount of paper created as media of exchange for these things multiplied. Law's ideas eventually proved correct. Paper finance had a life independent of moneybags in a cellar. The dynamic growth in manufacturing and trade was an engine of wealth in its own right and could only be practically represented by paper certificates agreed to have a certain meaning and value.

In the early days of the United States there were very few corporations, and of those almost all were created for the building of public works such as canals, roads, and bridges. Their officers could be held personally responsible for failures and disasters. Their charters lasted between ten and forty years, often requiring the termination of the corporation on completion of a specific task. In the 1840s, as railroads began to be organized, the nature of corporations changed. Railroads certainly functioned as public amenities but, unlike canals, they were organized as private money-making ventures. Factories, too, began to organize on a scale much larger than the individual workshop. Technologic innovation prompted the need for corporations to define their own purposes, not have one imposed by the government.

By the 1850s, the idea of limited liability began to be adopted in law. Officers of corporations were no longer held personally liable for the financial vicissitudes of a venture, apart from cases of criminal wrongdoing—and there was broad latitude in this, too, if only because the law lagged behind new swindles being innovated alongside new technologies. Under limited liability, a corporation could go bankrupt, but the personal assets of its executives and stockholders enjoyed protection. A corporation could be sued for some misfeasance, and perhaps ruined, put out of business, but the officers were not necessarily subject to civil damages. By 1886, the U.S. Supreme Court decided that corporations essentially had to be treated as "natural persons" under the law, specifically the fourteenth amendment to the Constitution, which had been crafted recently to protect freed slaves in the post-Civil War South. A corporation was able to use this new "identity" as a means to escape onerous regulations that might abridge its life, liberty, or property. Finally, the life of this fictitious corporate "person" was no longer deemed to be limited to any term of years but would be permitted a kind of immortality, to continue on past the lives of its founders.

The emergence of the modern corporation, along with new industrial technologies and the increased energy inputs of fossil fuels, led to the economic free-for-all of the late nineteenth century. Great complex ventures such as Standard Oil and the Union Pacific Railroad rose, financed by the issuance of shares. There were problems with this new way of doing things. All-out competition between companies in a given field tended to resolve in monopolies, and rather swiftly, too. There were many opportunities for mischief among the corporate officers, such as the "watering" of stock and the cornering of markets in commodities, leading some industrialists to be called "robber barons." Means had to be devised for regulating the immense amount of tradable paper generated; these were institutionalized in banks and stock exchange protocols. Standards and norms of operation for the trading of paper "assets" were established so that the public could agree on the value of things in order to trade them fairly. People got used to the idea of stocks (shares in a company) and bonds (units of debt owed by a company or government at interest) as elements of daily life—at least well-off city people did—and these instruments became normative devices for managing surplus capital, i.e., wealth. Skepticism about the reality of these items persisted, especially among the large rural population, and was reinforced by a business cycle that remorselessly went bust at intervals, leaving families wrecked as if they had been hit by tornadoes, and shaking the very consensus of hopeful expectation that underlay the acceptance of abstract finance in the first place.

What sustained fundamental faith in all these novelties of finance and capital was the continually upward-ratcheting industrial growth despite periodic reversals, which was made possible by the constant increase in available energy, that is, fossil fuels. In America particularly, offering surplus ecological carrying capacity to Europe's saturated habitats, a massive wave of immigration between 1880 and 1920 sustained the idea that growth was a permanent feature of the modern economic landscape. The business cycle might go boom and bust, but when the next boom occurred, there would always be more. More growth. More available energy. More commodities. More finished goods. More grain and beef. More immigrants coming from the constrained ecologies of Europe. More demand for things. More jobs. More production.

In the period between the end of the Civil War (1865) and the outbreak of World War I (1914), the middle classes continued to expand and to enjoy ever greater material comfort as technology delivered one new wonder after another: indoor plumbing, painless dentistry, the telephone, electric lights, motion pictures, cars, flying machines. The first decade of the twentieth century represented in many ways the summer of industrial civilization and the capitalism that served it. The word "progressive" is associated with American politics of those years, but the idea of progress in all aspects of life saturated the spirit of the time. The period just before World War I capped a long period of relative peace. (The Spanish-American War was inconsequential compared with the Civil War.) Giant corporations, tamed by new mechanisms of law and regulation, seemed to be reaching higher levels of efficiency, rectitude, and respectability—despite occasional calamities such as the Pullman strike or the Triangle Shirtwaist fire. These corporations would deliver the good things of life, while muscular government under Teddy Roosevelt would make sure that they behaved themselves. Trade unions gained "a place at the table." Public schools effectively converted hordes of foreigners into grateful new citizens who could read, do long division, and even advance socially to vertiginous rungs of society. The big cities bloomed in the great Beaux Arts projects of civic beautification. Public libraries and museums opened in grand new buildings and anybody could go into them. Public health and sanitation improved markedly, especially as cars rapidly began to replace the horse (the Model T Ford was introduced in 1907). Outright magic was being conjured in the form of motion pictures and in wireless radio that could transmit the human voice vast distances through thin air. The human race seemed to be entering a golden age.

It is therefore hard to overstate the devastating effect of World War I on the psychology of the Western industrial nations. The public consciousness was not prepared for the industrialization of slaughter and the war quashed the techno-optimism of the Belle Epoch. After 1918, Europe entered a funk of imperial decline, contested authority, and terrible turbulence in culture and politics. Three great dynasties—the Hapsburgs, the Hohenzollerns, and the Romanovs—had been toppled. Bolshevism was starting an experiment in Russia that would lead to the greatest bureaucratic murder spree in human history. Germany slouched toward bankruptcy and political psychosis.

In America, which was physically removed from the scene of battle and suffered fewer combat deaths than the Europeans, the aftermath of World War I was not a funk but a decade-long mania that turned the optimism of the Progressive Era into a fugue of sensation-seeking and escalating unreality that terminated in the economic nervous breakdown of the Great Depression.

Industrialism and Entropy

The financial frenzy of the 1920s was a mania stoked by oil, and specifically an economy intoxicated by automobiles and the first great wave of suburban expansion—all of which generated immense business activity in everything from land development to appliance manufacturing. The goods of mass production diffused through America with astonishing speed. Eight percent of American households were wired for electricity in 1907; 35 percent were wired by 1920. Car production shot up from 45,000 units in 1907 to 3.5 million in 1923. Most importantly, all of the nation's oil needs were met by oil easily obtained from safe places within U.S. borders and cheaply distributed all over the country on the world's finest railroad system. Public money was redirected from the Beaux Arts projects of civic beautification into retrofitting the cities for cars, and into paving roads out from the cities into the new suburban hinterlands. An epochal transformation appeared to be under way and the opportunities for growth and fabulous profits seemed limitless. Having made the world safe for democracy, Americans now felt entitled to benefit financially from its marvelous workings. The Progressive Era optimism about the quality of civilization and the public virtues associated with it mutated into a widespread lust for private riches and private luxury, especially among the middle classes. Privatizing amenity is what the new life of suburbia was all about. Stocks issued by companies making all the furnishings of suburbia, such as radios and cars, seemed devoid of hazard and the middle classes dove giddily into new investment pools, the precursors of mutual funds. All one needed was 10 percent down, and a double in the price of the shares would multiply the stake tenfold. By the end of 1929, total dollar trading volume in paper securities had reached 133 percent of gross domestic product. While the twenties roared, on Wall Street there were ominous rumblings in the background of the real economy.

For instance, the same oil-fueled boom that energized the suburban expansion of the 1920s brought turmoil and trouble to the farm economy. Thirty percent of the U.S. population still lived on farms in the 1920s. U.S. farmers had done well during World War I, exporting grain to a Europe that had become a shell-blasted battlefield. By the early 1920s, though, Europeans were able to feed themselves again. Meanwhile, the introduction of the tractor and the mechanization of farming in the United States led quickly to massive overproduction of grain. Unable any longer to pawn off the surplus on Europe, America suffered a crash in grain prices. The farm depression, which preceded the financial depression by half a decade, was a self-reinforcing feedback loop. As the market prices of corn and wheat plunged, farmers desperately tried to make up for low prices by producing more, which the domestic markets could not absorb, leading to even greater surpluses and more depressed prices.

Farmers had quickly become addicted to a new debt system of annual operation, mortgaging their farms to raise cash to pay for new machinery and fertilizer—literally betting the farm on a good crop. With prices chronically depressed, mortgages could not be paid off. Farm foreclosures soared in the mid-1920s. Another unanticipated consequence of mechanized farming was the destruction of soil. The tractor and its implements were machines that no one had previously experienced before, and it was some time before farmers noticed the insidious effects of soil compaction, rutting, and erosion that occurred. This would combine a few years later with an extended drought to produce the additional hardship of the Dust Bowl.

Strange things were also happening in the manufacturing sector, supposedly the nation's great strength. Markets were becoming saturated. By the late 1920s it was no longer possible to sell as many electric hair curlers, alarm clocks, and especially cars as American factories could turn out. Everyone who could buy a car had bought a car. Henry Nash had warned his fellow carmakers in 1925 that their markets were approaching saturation, and they had laughed at him. They were intoxicated with the steady, stupendous growth in annual sales they enjoyed since the end of the war and they were sure the boom for cars would keep going. But other conditions were changing beyond the walls of their tunnel vision. By the mid-1920s, the great wave of immigration suddenly ended. The National Origins Act of 1924 and other measures set new highly restrictive immigration quotas that cut new admissions to 2 percent of each nationality from the 1890 census. This choked off what had been a constant half-century-long demographic subsidy of ever more customers for U.S. manufacturers. At the same time, foreign markets for cars were very hard to penetrate. Europeans had little oil of their own, paid a lot more for what they could get from far away, were exhausted financially from the war, and had land tenure practices that did not encourage suburbanization. Besides, they had their own carmakers for their own limited markets. Other impediments to trade were created by politics. In 1922, President Warren G. Harding had signed the Fordney-McCumber Tariff Act into law. He praised it as one of the greatest tariff acts in the history of the United States and declared that it would contribute to the already-growing prosperity in the nation. Rising tariff barriers in the United States made it more difficult for European nations to conduct trade with the United States and, as a result, to pay off their war debts, which caused additional strains in the financial markets. It also gave them incentive to erect retaliatory tariffs of their own against American-made goods and commodities. Meanwhile, the Asian market for cars and electric appliances was insignificant. Most Chinese lived in the equivalent of medieval conditions. Drivable roads barely existed from Shanghai to Istanbul. Most of Asia was still unelectrified.

By the late 1920s, the sheer expectation of ever-rising stocks trumped the reality of the production and trade problems lurking in the American economy, and investment turned into mad speculation. Finance came to be viewed as a productive activity itself rather than a means to promote production. The public was no longer buying stock to invest in enterprises that would pay dividends over time, but merely because one could get rich from buying and selling stocks. As more people bought in, stock prices climbed still higher—a dangerous positive feedback loop. During the height of the mania, of course, many more stocks went up then went down. The middle class entered fearlessly into the realm of the "sure thing" reassured by the presence of the new and august Federal Reserve as a force that would stabilize the business cycle, and by such authorities as Harvard economics professor Irving Fisher, who declared the arrival of permanent prosperity a few weeks before the stock market crash of October 1929. (Fisher had earlier written a book saying that Prohibition would direct more dollars formerly spent on liquor into "home furnishings, automobiles, musical instruments, radio, travel, amusements, insurance, education, books and magazines.")

A great aura of mystery still surrounds the Great Depression that gathered momentum after the stock market crash of 1929. How could such a calamity happen amidst such plenty, Franklin D. Roosevelt asked rhetorically early in his presidency. It was indeed a conundrum. America was flush with oil, coal, wheat, ores, timber, and underutilized manufacturing capacity. There was plenty of stuff to be had, but suddenly too few had money to trade for stuff. There are technical explanations such as Milton Friedman's idea that it was a normal recession made infinitely worse by foolish government policy, or by economic leftists who maintained that more resolute government spending would have helped the "normal" business down-cycle correct itself more efficiently. What we know is that the stock market crash set in motion another powerful and destructive feedback loop. A lot of overvalued stock had been purchased on credit. When the share prices of many stocks collapsed, banks were left holding collateral in stock worth a great deal less than the money they had lent out to people who bought the stock. The stock market was hardly the only thing that brought down the banks—collapsing asset value in real estate and other collateral played its part—but by 1934, 11,000 banks had failed or had to merge, reducing the number by 40 percent, from 25,000 to 14,000.

Essentially this led to a widespread loss of faith—of consensual agreement about the nature of what modern credit had become, and of finance associated with the idea of credit (of "money" created out of nothing more than expectation), and of all abstract instruments of finance that had accumulated in daily practice over a period of about a century, such as stocks and bonds, to service these abstract notions of value. Loss of faith in credit dried up credit, as credit was based on little more than faith. The industrial economy ran on credit, most especially its "normal" expected annual "growth." The shattered consensus over the value of money led to dangerous fluctuations, in one nation and another, that we call inflation and deflation. There were many technical explanations for these phenomena, too, some having to do with the pegging of currency values to the more tangible value of gold—but the trend for two hundred years, since the time of John Law, had been to set credit free of its limiting association with bags of gold in a cellar, and this idea had been reinforced since then by economic theorists from Adam Smith to David Ricardo to Alexander Hamilton to John Stuart Mill.

Whatever was considered "normal" about industrial-age finance in 1929 was more a collective fantasy about an unresolved social experiment than a set of established facts of ecological economics. The industrial "story" as a phase of human history was barely two hundred years old, and it was a radical departure from the way human beings had done things for fifty thousand years, or even during Ben Franklin's childhood. In the early twentieth century, the mechanized factory system was new, mass production was new, broad-based installment purchases were new, the thirty-year mortgage was new. Modern economic behavior was being improvised. Not even savvy individuals knew what to expect, only what they hoped for. Except among a few intellectual eccentrics, such as Oswald Spengler, there wasn't any notion that the industrial story had a beginning, a middle, and an end, or where we were in the story. Nobody before had ever seen anything like a runaway stock market at this order of magnitude, or one so complexly integrated with international economies, dependent on imagined ideas of value, as had developed by the early twentieth century. Nobody had ever really seen this combination of circumstances. It was all a novelty.

One thing was clear after the debacle of World War I. Virtually all the industrial economies were sliding into terrible trouble and their trouble was all based, in one way or another, on the distortions that grew out of the divorce between ecological economics and an economics of abstract finance—otherwise known as capitalism. Some of the remedies for the ills of the capitalist experiment were worse than the experiment itself, most particularly communism, which purported to end inequality of income by abolishing wealth (and private property), and which succeeded only in making everyone poor—everyone, that is, who survived the bureaucratized mass killings instigated by the lawless gang that ran the communist party of the Soviet Union.

Germany, which had suffered terrible economic hardship after World War I, climbed out of the Great Depression relatively early in the 1930s, but only because Adolf Hitler so militarized and bureaucratized the nation that it became, in a very few years, a kind of meta-machine in which people were reduced to interchangeable parts (or disposed of if they didn't conform and perform).

In America, after the crash of 1929, the loss of faith in various forms of credit represented by abstract instruments of finance translated into a persistent lack of money—that is, a means of exchange—and the institutions devised to create it stood in disrepute. People could buy very little. Business stagnated. Companies would not hire workers when there was so little demand for products. It was a vicious cycle and it had vicious side effects. Another way of looking at the financial debacle of the 1930s is an ecological view such as William Catton's metaphor of the industrial economy as a "detritus ecosystem."[3]

Catton argues that the human race living off the "drawdown" of nonrenewable fossil fuel resources is the equivalent of the algae in a pond enjoying a temporary rush of nutrients in one brief season. Catton's analogy can be applied and extended to clarify the Great Depression in the context of ecological economics. After the crash of 1929, something had definitely changed in America. But the puzzling part is that the "nutrients" in the form of cheap oil—the "plenty" Roosevelt spoke of—still flowed. So why did the economic environment become so intractably unhealthy? From an ecological view, the Great Depression represented the effects of severe socioeconomic "pollution" produced by the oil-fueled boom of the 1920s, and this "pollution" had the effect of "poisoning" the financial ecosystem and consequently killing off financial "organs" that people had come to depend on in order to "thrive" (i.e., to grow wealthy and reproduce). Specifically, the "pollution" killed off the

3. William Catton, *Overshoot*, Urbana and Chicago: University of Illinois Press, 1980.

organs that generated credit and turned it into money. This systemic "pollution" of the financial ecosystem harmed the industrial environment enough to temporarily quash any further exuberant "growth." There was no human die-off but there was a die-off of expectations and a reduction in carrying capacity of the U.S. economy.

Is it fair to say that the by-product of zealous oil use literally converts into such an abstract form of "pollution" capable of poisoning what amounts to a social consensus? This must return us to the idea of entropy. Entropy is the spending down of energy and its translation into negative by-products. Not all of them are physical or material. Air pollution is one expression of entropy. But so is social disorder. So is institutional breakdown. Bodily death is another. These negative by-products of entropy can become interchangeable as entropy progresses, depending on any combination of variable conditions and circumstances. A careful reading of twentieth-century history would bear this out. In the modern era, entropy has been expressed in conditions as seemingly unrelated as war, industrial pollution, pornography, mass political murder, the shattering of a consensus about the value of money, and incompetent parenting. The introduction of high entropy into a given system is profoundly destabilizing in many ways. Entropy, like God, moves in mysterious ways.

For instance, to many historians the precise cause of World War I remains an abiding puzzle. Why should the assassination of an Austrian prince in Serbia lead England and France to yield a combined 2.2 million battlefield deaths as a result of the war following that event? What did England and France really care about Serbia, or for the Hapsburg royal family? Austria was already in steep decline as a political power. Why did the Russians eventually give up an estimated 1.7 million lives in this struggle that involved no vital interest, and then turn on themselves in a brutal civil war yielding a gangster-style dictatorship that brought on even more wholesale death? There have been many explanations. Most focused on the abstruse diplomatic machinations of the day, and they are all more or less inadequate. No territory was really at stake in World War I, at least in Europe proper, where the war was fought, and preceding the war there had been no significant friction in the far-flung colonies owned by the great powers. Was the "honor" of a few diplomatic alliances worth so many lives?

I think only an ecological explanation will suffice. World War I happened just as the industrial nations had entered a crucial transition from the coal phase into the oil phase of the industrial narrative. The human race was in the process of ratcheting up from one level of high-entropy activity to a yet higher one, meaning that there would be many more by-products of increased entropy as oil came into greater use. It is interesting to note that beginning in 1911, several years preceding the war, Winston Churchill, then lord of the admiralty, worked feverishly to convert the British fleet from coal to oil power. Oil-powered warships were more powerful and had far greater range than coal-fired ones. Note too, that the opening campaigns of the war were carried out using the techniques and logistics of the previous century—trains carried soldiers to the front, and millions of horses and mules were engaged to haul the artillery and supplies—but within a short time the combatants had shifted to gasoline-powered automobiles and tanks. Guns and artillery, too, operated at higher energy coefficients and killed soldiers in greater numbers than in previous wars. The sudden increase in these and other energy discharges by the great nations led quite naturally to an increase in entropic by-products, namely disorder, environmental destruction, and death.

Further entropic by-products included the "death" of optimism about the coming golden age of the twentieth century, something with more than abstract intellectual ramifications. That led to the now mythic disillusionment with civilization that followed the war, the loss of faith in institutions, traditions, and authorities (manifesting in part as the deliberate disorders of Modernism in the arts), and the profound collapse of the German economy under the terms of the treaty of Versailles. The particular circumstances of America after World War I were different, as we have seen. Germany sank, America boomed. For a while.

An ecological view of history could interpret the rise of totalitarian government as yet another by-product of high entropy. Both Nazism and Soviet-style communism might be described as politics "polluted" by insane ideology—a consensus disorder, often characterized as mass psychosis. Both systems grew out of social distress provoked by industrialism. Both systems undertook the extreme regimentation of their citizens as a defense against disorder—against entropy. The logic of the machine was overlaid on all social relations at a scale identical to the mass production of factory goods, and by similar methods. In the process, these systems achieved unprecedented industrial efficiencies in killing off those citizens unsuitable for regimentation. Stalin's terror and Hitler's holocaust were regimented die-offs. Adjustments to ecological carrying capacity were carried out with the remorseless logic of Taylorism.[4] World War II was an additional industrially organized die-off, with accompanying massive environmental destruction and social disorder. When it was over, the European principals were battered and entropically wasted.

America had participated in the military die-off of World War II to the extent of 295,000 killed in action, but its industrial engines of production and entropy creation remained intact, along with its reserves of oil and the infrastructure for producing it. After the war, the United States embarked on the high-entropy projects of building a suburban drive-in utopia and a nuclear arsenal. The first was a living arrangement with no future, and the second was the ultimate expression of entropy—an industrial means for sterilizing the planet Earth of all life.

The Entropy Express

In the evolving story of the industrial era, abstract notions of monetary value oscillate between the condition of comporting with reality to being wildly at odds with reality. A consensus about monetary values that align with reality (a period of growth and social stability) morphs into a fantasy in which values diverge from reality, and then a catastrophe like the crash of 1929 occurs to correct the fantasy (often inducing social disorder) as the collective imagination struggles to get values back in line with reality. This is another way of describing boom and depression. Emotion tends to govern events, which is one of the reasons that economics resists all attempts to rigorously empiricize it. In a depression, the loss of faith can become just as inordinate as the mania of overconfidence in a boom.

4. Frederick Taylor wrote his supremely influential book, *The Principles of Scientific Management*, in 1911.

During the Great Depression, the United States still had plenty of oil available to nourish industrial activities. But the consensus about the value of paper assets had been shattered by the collapse of asset prices in 1929 and the ensuing bank failures. The fear of capital markets grossly misbehaving had become so extreme that the faith they operated on would not re-form and capital (in the form of credit) could not be raised for industrial enterprise, despite the "plenty" of raw resources and manpower. Without enterprise there were fewer jobs, fewer paychecks, less demand for enterprise, and so on, a vicious downward cycle. The onset of World War II finally yanked America out of this self-reinforcing feedback loop. In a military emergency, wallowing in a state of economic anomie was no longer an option. The consensus was compelled to re-form, and the new consensus was centered on the idea that the government would become the prime customer for manufactured goods (i.e., war materials), using dollars "issued" by the customer itself, which is to say government-backed credit. This led quickly to a robust circulation of cash money—the very thing that had been so conspicuously absent during the depression.

Militarizing the economy was as effective a tonic for America as it had been for Germany nine years earlier. Car factories now turned out airplanes and tanks and operated at full capacity. As able-bodied men were drawn into military service, farm labor went from a state of depressed surplus to one of critical shortage. Farm worker pay in the early 1940s shot up from a dollar a day to a dollar an hour. Women were "drafted" into the realm of heavy factory labor, previously the sole domain of males. Superior production capacity, undisturbed by bombings, was largely responsible for America's victory—along with enough domestically produced oil to fuel the military mega-machine.

The entropy produced in World War II was much more widespread and profound than that of World War I. In World War I the action had taken place almost entirely on rural terrain, classic battlefields. In World War II, much of the warfare was urban. The long-range bomber had reached a high stage of refinement in the twenty-plus years between world wars. None of the major capitals had been damaged in World War I. In World War II, hundreds of towns and cities were destroyed in Europe and Asia. Berlin was reduced to gravel; London was badly mutilated; and, of course, Hiroshima and Nagasaki became radioactive ashtrays. The casualties of World War I had been enormous, astonishing, appalling beyond civilized peoples' wildest dreams, but the victims had been overwhelmingly soldiers. The casualties in World War II were overwhelmingly civilians and in much greater aggregate numbers.

Through the 1950s and 1960s, Europe, Japan, and Russia succeeded in eventually resuming industrial activity. The fear of atomic weapons constrained overt conflict between the United States and its chief ideological adversary among these recovering industrial powers, the Soviet Union. The wars of this era would be proxy wars, fought at the smaller scale, using mainly older weapons and tactics (certainly not nuclear missiles). Following demilitarization in the late 1940s the United States swooned back into economic malaise. The memory of the Great Depression was still vivid. But conditions had changed. All the barriers to world trade were down. With its industrial capacity still intact, and everybody else's wrecked, the United States owned world markets for factory goods. Confidence from winning the war now extended into other spheres of American life. A new generation of financial leaders realized that not only could credit and money be "created" pretty much at will to make things happen, but we could also extend credit to the exhausted powers who had been our allies and enemies during the war and get them to buy the products made here, too. It was surely more productive than heaping onerous reparations payments on them, which had led previously only to Hitler and more war. The United States, as the remaining noncommunist great power—i.e., the one that believed in money, credit, and finance, per se—was able to restructure a world financial system that would assure, at least, maximum stability if not prosperity. This was formalized in the Bretton Woods agreements, drawn up in a series of meetings before the war had actually ended.

The common view among the participants was that the worldwide depression of the 1930s and the rise of fascism could be traced to the collapse of international trade and isolationist economic policies. The aim of Bretton Woods was a planned global regulatory framework for trade and finance, establishing a postwar international system of convertible currencies, fixed exchange rates, and free trade, with the American dollar as the benchmark for all relative values. Institutions were created to regulate these agreements: the International Monetary Fund, the General Agreement on Tariffs and Trade (GATT), and the World Bank (formally called the International Bank for Reconstruction and Development). An initial loan of $250 million to France in 1947 was the World Bank's first act. The idea was to create a credible structure for international confidence to dwell in, so that investment could take place in a world traumatized by war, depression, and more war.

It worked pretty well for about thirty years. During that period, America turned back to the suburban expansion project—started in the 1920s—as a replacement for military mobilization in reinvigorating the economy. The suburban project seemed to make sense. It would enlist all the great industries of the nation—steel, concrete, the building trades, the carmakers, the appliance manufacturers, the realtors—and give them plenty to do. In 1950 we had lots of oil, reasonable expectations of discovering more, and no sense of limits. World oil demand was still very modest. China, India, Africa, and South America used relatively little oil, and war-enfeebled Europe only slightly more. Americans felt no qualms about using as much as they wanted.

As a social project, suburbia made even more sense after World War II than it had in the 1920s. For one thing, the cities themselves were in even worse shape after twenty years of depression and war. They were grim, industrial hulks, with outmoded row houses, tenements, and dreary streets now dominated by cars. Few new buildings had gone up after the crash of 1929, especially housing, and the old ones had had their maintenance deferred. The infrastructure was old and tired. Everything about our cities reinforced the traditional American antipathy toward city living (and its corollary that country life was the antidote).

The rural hinterlands were full of cheap developable real estate. Men such as William Levitt brought military production methods to the task of suburban house construction and new subdivisions such as Levittown were an instant sensation. That they were a cartoon of country living rather than the real thing didn't seem to bother the buyers, for whom anything other than a canebrake in the Solomon Islands or an apartment at eyeball level with an elevated subway was an improvement. The new suburban houses themselves may have been dreary little boxes under a thousand square feet in their own right. But each had its little patch of green lawn, and volumes of fresh air surrounding them, and most of all they were not in the city, nor contaminated by the proximity of obnoxious, smoky, noisome city-type activities having to do with trade or manufacturing. Thus the stage was set for the postwar economy.

That economy featured, most importantly, the rehabilitation of finance. American life, with its twin engines of suburbanization and factory production of consumer goods for the whole world, became so quickly and obviously successful that a new consensus formed supporting the value of the dollar and its paper accessories in capital markets, chiefly stocks and bonds. This is not to say that the securities markets boomed in the 1950s and 1960s—it took until then just to recover the value levels of the pre-1929 crash—but stocks and bonds did regain respectability, legitimacy. Those who had lived through the Great Depression, meaning virtually all the men who had served in the wartime army, had very modest expectations about the role of finance in the postwar economy. In the 1950s and 1960s, Americans bought stocks for the annual dividends they paid, not to flip them for a quick profit. In fact, share prices remained relatively very flat during this period. The whole notion of investment was different than it would become later in the twentieth century. In the 1950s and 1960s, stock and bond values were linked much more directly with the successful production of real goods. General Motors derived its profits and paid its dividends on the basis of auto sales, not as today, primarily from leveraging interest rates and other abstract numbers games removed from the actual making of products. In sum, the public attitude about the role of finance was extremely conservative. Finance was not an "industry" per se, but a set of institutions designed to keep the idea of money and its accessories credible, so as to allow real industries to function. A small fraction of the public bought securities, a tiny fraction of the public actually made their livings in finance, and the majority of this tiny fraction—the workaday stockbrokers, bankers, commodities traders, and so on—had incomes that would seem laughable by today's standards. They were middle-class.

Indeed, the middle class in America was never broader. Differences in pay scales from the very top to the bottom in American life at the mid-twentieth century were amazingly democratic by today's standards. From 1947 to 1968, the wage inequality between top executives and lowly workers actually dropped steadily. In the 1960s, automobile assembly line workers made more money than college professors. In 1960, the pay of company CEOs was on average forty-one times the pay of the company workers; by the year 2000 the multiple for CEO pay reached 531 times the pay of a worker.

Banking also regained respectability after the calamities of the 1930s. Federal deposit insurance, which had been instituted in the depths of the Great Depression, and only for deposits under $2,500, was raised to $10,000 in 1950, and the middle class was induced to feel confident about keeping its money in banks again. Interest rates remained modest, but so did inflation. The influx of savings made money available in capital markets to invest in new ventures. It was real money derived from work already done, pay already earned, true capital. Before the great orgy of mergers and consolidation that began in the 1970s, retail banking was largely local and community-centered. Bankers made loan decisions based on firsthand knowledge of projects going on in their communities—not, as today, based on bundling and selling clumps of mortgages for generic suburban developments they have never laid eyes on.

The baby boom generation, the offspring of those who fought in World War II, grew up in this period of extraordinary financial stability and economic promise, and it became their lifelong benchmark for normality. Other bogeymen lurked in the shadows of American culture during the Eisenhower years—nuclear war, racial inequality, Sputnik—but few Americans doubted the soundness of the dollar or the sanctity of the New York Stock Exchange. In fact, the consensus about the rightness of the U.S. economy was so broad and sturdy that the baby boomers revolted violently against its chief manifestation, belief in the value of money, as soon as they entered adolescence. The assassination of John F. Kennedy in 1963 was certainly a crisis point for the collective boomer psyche, since it shattered virtually all of the shared sense of security about the rightness of American political and economic life. The death of JFK sent pubescent baby boomers into a deep funk, from which they emerged with a new and rather strange worldview.

Childhood's End

The rebellion of the hippies—of which the author was nominally one—based itself on the notion that abundance was a natural entitlement and one could "drop out" of an insecure, deadly, and frightening industrial culture to live off the fat of the land. It was inescapably a jejune philosophy, fraught with contradiction. For the hippies the natural order of things included items such as stereo record players, electric guitars, motor vehicles for adventuring around the country, cheap bulk whole grains, and other products of an oil-intensive industrial way of life. The hippie platform, so to speak, with all its mystical incunabula, rested on the platform of "normal" American life, and would have been impossible without it. The Vietnam War certainly intensified the revolt by threatening to send a cosseted generation off to slaughter for reasons abstract at best and absurd at worst. Thus, the suburban consumer economy, a.k.a. the American Way of Life, which underwrote the war, fell under a cloud of opprobrium along with the idea of money itself.

Interestingly, the hippie revolt entered oblivion at exactly the same time that real challenges to the economic status quo presented themselves. These challenges came in the hidden form of the U.S. oil production peak around 1970 and the effects emanating from it. The public was completely clueless about the U.S. oil peak for three years, until the OPEC embargo of 1973 made the connection for them, and even then only a tiny minority understood its significance. Between 1970, when the U.S. peak actually occurred, and 1973, when the OPEC embargo ushered in a new age of global oil politics, the effects of the U.S. peak were expressed solely in arcane issues of fiscal policy, which the public also generally did not understand. The U.S. government had rung up large deficits in simultaneously funding the Vietnam War and an expanding agenda of social programs. These caused additional distortions in the economy.

By the early 1970s, other national economies were fully back on their feet and, led by Japan, were selling a lot of cheap exports to America, which were, of course, paid for in dollars. America's trade imbalances, aggravated by the increased importing of foreign oil, grew steadily worse. Overseas dollars were accumulating and could still be exchanged for gold under the Bretton Woods agreements (at the $35-an-ounce rate that had remained unchanged since the FDR days). Only now there were far too many dollars in other countries to be exchanged for gold at the official price. Eroding confidence in the dollar threatened mass redemptions of American gold. Foreign governments, or their central banks, could show up at any time at the "gold window" of the U.S. Treasury and insist on trading in their dollars for gold, which would precipitate a run. In fact, in the summer of 1971, the British ambassador formally requested $3 billion in bullion from the U.S. Treasury. That August, President Nixon closed the gold window, effectively detaching the dollar from anything but an abstract notion of its value. At that point for the first time in history, formal links between the major world currencies and real commodities were severed. As the price of gold levitated up toward $140 an ounce, the foreign oil producers, especially Arabs spooked by the dissociation of money from gold, sought to raise the dollar price of their crucial commodity.

The Watergate scandal effectively distracted the American public from the momentous transition going on in global economic relations. While Nixon tergiversated, the United States quietly surrendered its position as the world's swing oil producer, meaning that it lost pricing control of the world's single most crucial commodity. The price of oil was on its way up well before the OPEC embargo of 1973 (see Chapter Three for the particulars). The United States emerged from that trauma with its fiscal and monetary foundations shaken. Loss of pricing control over oil led quickly to the loss of control of all other prices, as oil was necessary for the production of all commodities and manufactured goods. America had never experienced such a loss of fundamental control over oil prices since the birth of the oil age in 1859, and it was unprepared for the consequences. Soon, the nation was gripped by rampaging inflation, surging interest rates, rising unemployment, and paralysis in productive activity. "Stagflation" appeared to baffle the conventional economists, but it was obvious that the interruption of utterly reliable supplies of cheap oil had queered the self-organized hypersystems that Americans had come to depend on for daily life—namely, all the motor-vehicle-based social and economic relations of dispersed living and production in a suburban nation. All of a sudden, in a world of expensive gasoline, 55-mph speed limits, and truckers' rebellions, the drive-in utopia sponsored by the Big Three automakers and their vassals wasn't working so well. The automobile industry itself was extremely hard-hit, as it was tooled up only to produce big "gas guzzlers" and the public suddenly developed a passion for much smaller cars—for which Japan and Europe happened to be already prepared. It would be almost a decade before Detroit answered with small cars of its own, and by then it had lost both market share *and* quality control. During the mid- to late 1970s, millions of individuals put off decisions to buy houses further away from their jobs, leaving the building trades in a profound funk. Meanwhile, other U.S. industries, including steel, textiles, and electronics, followed in the steps of the auto industry and began surrendering to overseas producers. Before the United States could regain its footing, it was smacked with another oil crisis, this one emanating out of the fall of the shah of Iran in 1979.

The American economy that emerged in the 1980s was battered. The first two years of Ronald Reagan's first term were dismal in terms of standard econometrics, job loss, and the like, but the world oil situation began to stabilize again, for two reasons: First the Camp David peace accords between Israel and Egypt engineered by President Carter had dampened the general level of worldwide Islamic enmity against the West—apart from the situation in Iran—allowing the Saudis to optimize OPEC outflows (and revenue inflows). Second, and perhaps more important, the coming on line of the Alaska North Slope oil fields, the North Sea fields, and deepwater wells that were the fruits of the desperate exploration carried out following the crisis of the seventies restored supply leverage to the non-OPEC nations. These discoveries, which appear now to be the last significant ones of the oil age, set the stage for the oil glut that characterized the 1980s and 1990s.

The oil glut had several ramifications. Foremost, it put the American public back to sleep on energy issues as they had been prior to 1973. They wrote off the traumas of the 1970s as a false alarm, a fake crisis cooked up by wicked oil companies and their Arab coconspirators, not what it really was: a preview of coming attractions. In any case, there was once again plenty of oil flowing, and in the final two decades of the twentieth century the oil only got cheaper and cheaper until it bottomed out at $10 a barrel around the year 2000. At the start of the oil glut, a climactic set of economic relations took shape led by Prime Minister Margaret Thatcher (and joined eagerly by President Reagan and his advisors) that would be called "globalism." It was not so much a new idea as the logical and inevitable result of mature self-organizing systems elaborating themselves under the influence of renewed, immense energy inputs—the ultimate cheap-oil way of doing business in the closed system that is the planet Earth. It entailed the maximization of short-term profit and the minimization of care for future generations. It was the ultimate generator of entropy.

The Final Fiesta

In America, globalism meant the accelerated dismantling of the nation's manufacturing base and its reassignment to other countries where labor was dirt cheap and environmental regulations did not apply. It also meant the ramping up of a "service economy" or, more properly, the myth of a service economy to replace the old manufacturing economy. I say "myth" because it was essentially absurd. It was like the old joke about the village that prospered because the inhabitants were all employed taking in each other's laundry. In fact, far fewer actual things of value were being created in the service economy. It was yet another temporary and protean manifestation of the tremendous entropy produced by inputs of cheap fossil fuel.

This wasn't the only myth, however. Another myth was something called the "digital economy." Computers came on the scene in a big way in the early 1980s, and by the mid-1980s the personal computer began to democratize the "information revolution." Computers changed a lot of things about the way business was done, but at the expense of enormous diminishing returns, which were rarely calculated into the dominant statistical analyses of our national condition. It was assumed, for instance, that computers greatly boosted productivity. Much of that gain was either illusory or fraught with collateral social and economic losses of other kinds. Companies that reported higher productivity were shedding employees like mad and the entire ethos of work in America was being transformed from one of people having secure careers and permanent positions with reliable companies to one of institutionalized insecurity for practically everyone below top management in a new general atmosphere of Darwinian corporate ruthlessness—under the rubric of "free-market competition." The computer revolution created an enormous structure of exploitation in the "service" sector of retail, for instance, from underpaid workers at the big-box stores to the supplier plants in China, where one factory was pitted against another to see who could fill Wal-Mart orders for less.

Another way of looking at the productivity myth was as a shifting of burdens from companies to the public. For example, most companies, government agencies, and schools computerized their phone systems, eliminating numbers of live human beings answering the phones. The net result was that it became nearly impossible to make contact with a live person at any company or institution in America. The public now had to waste astronomical amounts of time wading through tedious recorded phone menus or listening to Muzak while placed automatically on "hold," often getting disconnected, or ending up in voice-mail limbo. Communication was hampered by computerization, not facilitated by it. One of the obvious lessons was that human beings were actually better computers than computers. Human receptionists were much more adept than computers at evaluating requests and routing concerns in the proper direction. Under the new universal computer-managed regime, though, it was often impossible for customers to even order products from the companies that sold them.

The outfitting of corporate America with computer networks and systems for bookkeeping, inventory, shipping, and tracking certainly generated a lot of business and sales activity for the computer industry itself, and the boom of the 1990s was, of course, largely based on this tremendous installation of digital infrastructure and its regular updating every two or three years as the computers got more powerful. But that too was fraught with diminishing returns, and unanticipated consequences—another manifestation of entropy. The computerization of corporate America promoted the hemorrhaging of jobs and whole industries to offshore locations and the "outsourcing" of whole departments to other countries. Additional diminishing returns associated with the victory of national chain retail were the wholesale destruction of American communities, including both the "hardware" of towns and the "software" of social roles and networks associated with them. Computers only assisted predatory corporations in more successfully parasitizing existing value in victimized localities. They were most efficient at sucking the lifeblood out of complex communities. They helped "convert" complexity into simpleness (one big box instead of twenty-seven local businesses) and entropizing society.

Ultimately, the computer revolution led to the "dot-com economy" of the late 1990s, which amounted to a classic bubble over the perceived (or misperceived) moneymaking potential of the internet. A few gigantic successes were scored in Web-based businesses. Soon, investment banks were backing stock offerings on hundreds of businesses, many of which amounted only to a dream or a wish on paper. Vast amounts of money were raised in initial public offerings for laughable ventures, but the public had lost its critical faculties. Many investors knew nothing about computers anyway, or were intimidated by them. They had seen immense fortunes made by Microsoft, Apple, Oracle, Sun Microsystems, and the like. They even used Web-based businesses such as Google and eBay, and they assumed that some of the bright young dudes in black outfits and stylish eyeglasses behind the public offerings would be the next Bill Gates or Larry Ellison. Hundreds of other ventures were capitalized and geared up, and a stunning percentage of them failed. The diminishing returns of overinvestment had struck again. Entropy expressed itself in the form of mass delusion. The stock market, especially the tech sector, lost credibility, but there was still plenty of hallucinated wealth left in the American economy and, as we shall see later, it went somewhere else.

The Sprawl Economy and Funny Money

What one also saw in the America of the 1980s and 1990s was commoditization and conversion of public goods into private luxuries, the impoverishment of the civic realm, and, to put it bluntly, the rape of the landscape—a vast entropic enterprise that was the culminating phase of suburbia. The dirty secret of the American economy in the 1990s was that it was no longer about anything except the creation of suburban sprawl and the furnishing, accessorizing, and financing of it. It resembled the efficiency of cancer. Nothing else really mattered except building suburban houses, trading away the mortgages, selling the multiple cars needed by the inhabitants, upgrading the roads into commercial strip highways with all the necessary shopping infrastructure, and moving vast supplies of merchandise made in China for next to nothing to fill up those houses.

The economy of suburban sprawl was a systemic self-organizing response to the availability of inordinately cheap oil with ever-increasing entropy expressed in an ever-increasing variety of manifestations from the destruction of farmland to the decay of the cities, to widespread psychological depression, to the rash of school shooting sprees, to epidemic obesity. Americans didn't question the validity of the suburban sprawl economy. They accepted it at face value as the obvious logical outcome of their hopes and dreams and defended it viciously against criticism. They steadfastly ignored its salient characteristic: that it had no future either as an economy or as a living arrangement. Each further elaboration of the suburban system made it less likely to survive any change in conditions, most particularly any change in the equations of cheap oil.

It wasn't until the traumas of the 1970s that the finance sector mutated from being an adjunct of the industrial economy to becoming a leading "industry" in its own right helping to "drive" the economy. Among the distortions and perversions engendered by the "stagflation" economy was the rise of corporate cannibalism in the form of "creative" mergers and acquisitions, specifically hostile takeovers, the aggressive use of voting stock shares to gain control of companies that did not wish to sell, with the subsequent filleting and sell-off of assets, and discarding of the bones and offal (employee payrolls and obligations, careers, livelihoods, communities). The business culture celebrated the "corporate buccaneers" who engaged in these shenanigans as superstars the way Andy Warhol had elevated junkies and drag queens to celebrity a decade earlier. Of course, many businesses would not have been vulnerable to takeover if the entire manufacturing sector had not been wobbling with a range of problems from antiquated plants and equipment (steel) to inadaptable management (cars) to dismal quality control (electronics). The truth was that by the mid-1970s, American industry was uniformly showing signs of fatigue.

Banking underwent radical consolidation and change, too, following the disorders of the 1973 oil embargo and the Iran price-jacking of 1979. Big banks began gobbling up small banks, which had the effect of hypercommoditizing credit, turning it into just another "consumer" activity carried on at mass scale. Loans became ever more abstract "units" of generic "product," such as commercial mortgages, traded in bundles and clumps like scrap metal. As local business and local ownership became irrelevant, so did local banking and local lending for local ventures. The hypercommoditizing of lending disconnected bankers from knowledge of the ventures they lent money for—just so many strip malls or condominiums—which also tended to reinforce the generic predictability of suburban development all over the country. But this insidious surrender of human judgment would also work in the collective public consciousness to further abstract the nature of assets from the meaning of value or money as a general proposition. The entropy in this kind of banking produced huge diminishing returns that eventually showed up as a landscape defaced by ugly, clownish buildings deployed in wastelands of parking, built by people who didn't care about the places they were exploiting.

Parallel to the consolidation of commercial banks was the deregulation of the savings-and-loan (S&L) sector. These special banks, or "thrifts," were first chartered as a means to provide for long-term home mortgages. Before Ronald Reagan took office, S&Ls had to keep at least 80 percent of their assets in home loans, by law. A typical S&L would offer 3 percent interest on money deposited with it, and make mortgage loans to homebuyers at 6 percent. The 3 percent difference or "spread" covered the bank officers' salaries, and paid building rent and owner profits. Obviously this required a stable currency. The severe inflation and interest rate hikes of the 1970s threatened to drive the thrifts out of business. In 1980, Congress began eliminating the interest rate ceilings on S&Ls, and simultaneously raised deposit insurance from $40,000 to $100,000 per account for S&Ls. The 1982 Garn-St. Germain Act allowed S&Ls to invest up to 40 percent of their money in ventures not related to housing. This freed the owners of the S&Ls to invest in any cockamamie scheme, routinely awarding their banks substantial "points" for lending large sums to construct shopping centers, malls, condo complexes, and so forth, things associated with suburban sprawl development. Often these projects were egregiously unnecessary or redundant, but still very profitable for the bankers and developers. The bankers kept the "points" regardless of whether the projects failed, and these fees could be substantial for developments in the $100 million range. The developers also made out on super-generous fee payments to companies they controlled for superintending construction.

If the projects did happen to fail, so much the better. They could be resold (with more "points" garnered). Accounting irregularities abounded, often involving the multiple resale of defaulted properties at inflated appraisal values to conceal previous losses. The more confusingly complex the deals were, the more resistant they were to oversight and the more beneficial they would be to the participants. It was an obvious racket. The orgy of fraud led to the abandonment of the notion, ever associated with credit and money, that foolish lending would be rewarded by failure and ruin. In the case of the S&Ls and their officers, failure and ruin were heavily rewarded with federal deposit insurance payouts. As a nice side racket, thrift bankers could stash their money in multiple accounts in their own banks, and when their banks failed, they enjoyed $100,000 payouts per account in federal deposit insurance. When the S&Ls went belly up, of course, multitudes of ordinary citizens whose savings were lost also had to be paid out in federal deposit insurance. The result was roughly a half-trillion-dollar tab for the U.S. Treasury. Only a few of the most blatant offenders went to jail when their frauds were discovered, and several senators and congressmen saw their careers ended. But the most remarkable aspect of the S&L debacle was that the American public hardly felt any pain over it. The federal deposit insurance payouts were all subsumed in the gigantic deficits rung up during the Reagan and George H. W. Bush administrations.

Of perhaps greater impact on the finance sector, and on the meaning of money generally, was a set of new speculative activities based on the trading of "creative" financial instruments largely inspired by computerization. Computers could calculate large arrays of variables in ways never before possible. Even if they could not really predict the direction of markets—because of markets' essential nonlinear nature—the computer did increase the number of ways to play markets (and to lift the level of abstraction of "money" ever further away from real value-producing activities). They also enabled money—or, more precisely, the electronic notions and markers of money—to be transferred around the world at the speed of light. This allowed many new opportunities for playing minuscule changes in currency valuations and interest spreads around the world. It led to the "creative" invention of innumerable new "derivatives," or instruments based on values derived from other markers of value, contracts, or bets on the prospect of changing value of—anything. Stocks, mortgages, interest differentials, weather. In short, it was a way of turning all risk, as defined in investment terms, into a casinolike panoply of betting options in a new global investment casino.

During the formative years of the computer revolution, some players assumed that they had super-slick formulas or equations for beating the odds. They also employed strategies for "hedging" their bets so that one potential losing position would be covered by a winning position somewhere else. When extremely large figures were bet, even tiny value spreads could yield fantastic profits. It worked even better if the bets were leveraged—that is, if one had to put up only a fraction of the total bet in one's own cash—using money notionally "borrowed" from other sources to make up the rest, which was especially cool if the player was on a winning streak with snazzy hedging equations and the "borrowed" notional money never really entered the picture except as pixels on a computer. It was also a recipe for disaster. The poster child for the worst-case derivative fiasco was a hedge fund company called Long Term Capital Management (LTCM).

Hedge funds are largely unregulated on the assumption that their customers are wealthy, knowledgeable investors aware of the high risks and not in need of regulatory protection. LTCM was a kind of glorified and very high-toned "boiler room" operation run out of an anonymous corporate box building in suburban Connecticut, far from Wall Street. The firm was started in 1993 by a former vice chairman of Salomon Brothers (and champion bond trader), John Meriwether, who had had to resign from that investment house when an employee under his supervision was caught making false bids on U.S. Treasury auctions. So Meriwether went off with some of his most aggressive Salomon colleagues and opened his own shop far from lower Manhattan. The stars of LTCM's small staff of hotshot traders, computer nerds, econometricians, Ph.D. physicists, and math whizzes included two academic economists who won the Nobel Prize for their "contributions" to the understanding of option pricing—Myron Scholes and Robert Merton. Also on board was a former vice chairman of the Board of Governors of the Federal Reserve System, David Mullins. This crew of geniuses devised extremely complex mathematical models for hedged investment plays in global markets, mostly betting on variations in interest rates between U.S., Japanese, and European sovereign bonds. They would identify patterns in cycles of rising and falling rates and made their plays on the metatheory that markets invariably impose equilibrium on rates, which would revert to predictable norms. Within the range of differentials was a vast realm of small change that they figured no one else but LTCM could see, gazillions of "loose nickels that could be endlessly vacuumed up," in the words of Myron Scholes.

The enterprise, originally capitalized at $1.25 billion, had many leading banks among its clientele. Its moves were ultrasecretive. As a matter of policy, LTCM would not open its books or reveal its trading positions to even its best clients. Five years after its founding, it would control $134 billion in assets. For many of those halcyon years in the mid-1990s, LTCM showed annual returns in the 40 percent range. The principal partners were making scores of millions of dollars a year for themselves in vacuumed-up nickels. Merton and Scholes had such supernatural confidence in their own models that they calculated their chance of failure at zero in the lifetime of the universe and even over numerous repetitions of the universe.[5] In essence, they believed that computers combined with their own fabulous equations had given them the godlike power to completely eliminate risk in their business.

Disaster came rather sooner than the life of even this universe, in August 1998 to be exact, when the economic basket case, Russia, defaulted on its debt. With $4.8 billion in equity, LTCM had managed to leverage itself to the hilt by "borrowing" (in computer pixels) more than $125 billion from banks and securities firms and had entered into derivatives contracts (bets) with more than $1 trillion at stake. The Russian default set off a chain reaction of international flight away from low-quality sovereign bonds to U.S. Treasury certificates. This upset the reversion to predictable norms, or convergence, built into LTCM's models—interest rates were diverging wildly all of a sudden—which completely queered the immense bets they had placed on the spreads. In addition, LTCM happened to be in exactly the wrong place at the moment of the Russian default—reportedly 8 percent of its *book*, or $10 billion of LTCM's notional assets, were actually in Russian positions (bets). Russia, a gangster economy with

5. Roger Lowenstein, *When Genius Failed*, New York: Random House, 2000, p. 159.

nebulous property laws, shady banking, and little to no legal contract enforcement, was a risky investment, but if one had conquered risk, well. . . . Only now, instead of vacuuming up gazillions of nickels, LTCM was suddenly hemorrhaging hundreds of millions of dollars in collateral calls from their counterparties, those they had bets with.

With so much notional money at risk, in a world of money so abstracted from any real activity besides the trading of abstractions, the LTCM meltdown raised concern among leading bankers that the entire whirring skein of global digital (hallucinated) capital might unravel in a shitstorm of cross-defaults, leaving behind a lot of notional ruin—including the metanotion that any sort of financial paper issued by any nation or company had value. The chairman of the New York Federal Reserve Bank called in the heads of virtually every major bank in the city—all of which had lent LTCM piles of money to leverage their trades for fantastic profits—and in a matter of days the banks were persuaded to pony up several hundred million dollars each to recapitalize LTCM, under a new agreement that left the firm with only 10 percent of the action. Thus was LTCM saved from actually tanking and taking countless other international entities down with it.

The U.S. Treasury itself never actually contributed any money to the LTCM bailout. But the day after the bailout, September 29, 1998, Federal Reserve Chairman Alan Greenspan reduced interest rates a quarter of a point (25 basis points) to 5.25 percent, in the hopes of stabilizing the wobbling international bond market. He dropped them a second time a month later. From that point on Greenspan's Fed embarked on a long trail of lower interest rates—the magical generation of ever-easier credit—that spawned yet another episode of destructive mischief in the entropic economy: the real estate bubble, perhaps the last act in the sorry drama of the hallucinated economy.

Home: The Last Refuge of Value

It was perhaps natural that at a time when America had become a wasteland of traffic congestion and cartoon architecture, the public would invest such inordinate psychological capital in the idea of home ownership. The more degraded the civic realm became, the more the private realm mattered, because you could control it. The single-family suburban house had become the iconic symbol for several preoccupations of the collective national psyche: extreme notions of private property ("Don't tell me what to do with my land!"), the wish for security against "urban crime" (i.e., African American misbehavior), the wish for adequate public schooling, the longing for a peaceful sanctuary against the epic entropic disorders of "normal" late oil-age American life, and finally as a repository of family wealth in a time of financial turmoil. In the face of the things like the dot-com meltdown, the LTCM scare, the Enron scandal, and other disasters that eroded the notional value of financial paper, home ownership itself was now turned into a magical generator of unearned riches for both borrowers and lenders. It was consistent with the Las Vegas-ization of the national moral sense, chiefly the increasingly popular belief at every level of American life that it really was possible to get something for nothing. Anyone could see this in the easy public acceptance of gambling as okay and the proliferation of casinos everywhere in the land. Not even the evangelical Christians seemed to mind.

There is no such thing as intrinsic value in a house. A huge percentage of the public has now put its net worth into something that arguably isn't an investment. Apart from false econometrics of rising house valuations and the leverage that affords for raising cash within the context of the current lending rackets, a house is much more of a consumer product than an investment, especially the kind of houses built in recent decades in America, namely stapled-together boxes made of particle board and plastic cladding that require continual reinvestment in petty cash and labor for upkeep, and will probably not hold their value, even if well cared for, because of poor locational choices. A house on a one-acre lot in a subdivision in Loudoun County, Virginia, thirty-two miles from downtown Washington, may be a magnificent thing to behold today, with a soaring lawyer-foyer entrance, a restaurant-grade kitchen, and an inground pool out back. But if there is less gasoline to power up the fleet of cars necessary to service it, and no natural gas to heat the thousand-square-foot cathedral-ceilinged lawyer foyer, then chances are that the house is going to be a liability rather than an asset. When large numbers of house owners experience that transformation, the political fallout will also be transformative.

The house buying-and-selling orgy of the early twenty-first century was set off by the Federal Reserve's policy, over a five-year period from 1998 to 2003, of steadily reducing to nearly nothing the interest rate that it charged banks to borrow money, which worked its way through the lending system so that mortgage rates fell to historically supernatural lows. The low interest rates were joined by a further decay of lending practices so that practically anyone over age twenty-one with no record of creditworthiness could get a low or even zero down-payment mortgage. Other factors favored a flight of capital from other avenues of investment into houses. With interest rates under 2 percent, normal savings accounts and money market funds had become a joke. Economic pundits beat their breasts about America's pitifully low rate of savings—the conventional means for raising honest capital before the something-for-nothing fever seized the collective national imagination—but only chumps would save in passbook accounts at 1.75 percent. The dot-com meltdown had left a lot of the moneyed middle class feeling hosed by, and wary of, the equities sector. Perhaps even the deep resounding horror of the 9/11 attacks inspired a kind of bunker mentality that translated into a nesting mania. Wasn't that the appeal of Martha Stewart, the goddess of domesticity? So much of the surplus wealth remaining in America at the end of the twentieth century landed in the real estate sector under the theory that real estate was at least real. Finally there was the federal tax policy of the mortgage interest deduction that gave homeowners a substantial advantage over renters, which has always biased the U.S. market not only in terms of personal dwelling choices but in terms of housing typologies offered by the building industry.

True, the population of the United States was growing, but not at a rate that justified the construction of so many new McHouses, as the "units" were called in the pop-up subdivisions. Behind the phenomenal spurt of new construction was the still-accelerating flight not only from the cities, but also from the older suburbs, which were now infected by the spreading rot of the urban core. And propelling that spread was the fact that all through the late 1980s and 1990s, and into the new millennium, oil had only become cheaper in constant dollars until it stood at about $10 a barrel when the younger George Bush took office. This meant that, if nothing else, the nation could continue the suburban sprawl fiesta that had become the virtual replacement for the old manufacturing economy. It was, in short, another self-reinforcing feedback loop, a self-organizing system shaping the American landscape into a nightmarish diagram of motoring hyper-squalor. And underneath all of that was the credit creation machine of Alan Greenspan's Federal Reserve, manufacturing money electronically that wasn't really there, wasn't being accumulated through the traditional, and ultimately only, real means of savings on earnings from doing real work producing real things. It was a magic act.

The supernaturally low interest rates provoked an orgy of buying, and the orgy of buying bid up the prices of the houses, and as the prices of the houses levitated, the owners entered another new and strange zone of hallucinated wealth accumulation using the latest contrivance: the refinanced mortgage. Re-fi's allowed house owners to use their houses as though they were automatic teller machines. Say a person bought a house in 1999 for $250,000 and the house was appraised in 2003 at $400,000; that person could refinance with a substantial "cash out" privilege, converting the imagined increase of value into disposable income, which could then be used to buy motorboats, home theater plasma TV screens, or trips to Las Vegas. Refinancing prestidigitated an estimated $1.6 trillion for the American economy over a five-year period, and much of that "money" was deployed purchasing "consumer" goods—mostly made outside the United States. From 1999 to 2004 roughly a third of all house owners indulged in cash-out re-fi mortgages. The racket seemed without hazard when housing values only went up, up, up. Behind every extravagant cash extraction lay the belief that at some future date the house would be worth a lot more than the re-fi price and could be readily flipped. In super-hot markets such as the Boston suburbs or Long Island or Marin County, properties rarely stayed on the market for more than a few days. Often bidding wars broke out between hysterical buyers, going beyond the asking price.

Lending practices decayed further. New types of lending companies, such as Ditech, came along hawking "miracle loans" on TV with no closing costs, no down payment, making it possible for customers to sleepwalk into owning substantial properties or refinancing existing ones. Outfits like Ditech were a peculiar kind of financial animal, a mutant spawn of what previously had been known as the "sub-prime" market, meaning companies originally designed to serve high-risk borrowers, people with lousy credit records, deadbeats, bottom feeders, habitual bankrupts, *schnorrers*. After the mid-1990s, there was hardly a technical distinction to be made anymore between high-risk borrowers and everybody else in the casino atmosphere of America society. No one was at risk anymore because in the something-for-nothing economy it was impossible to be a loser. Or so went the herd thinking.

The decay of mortgage standards was abetted by the rise of the giant "government-sponsored entities" (GSEs), Fannie Mae (Federal National Mortgage Association) and Freddie Mac (Federal Home Mortgage Corporation). Fannie Mae started as a part of New Deal policy to stimulate the housing industry. In 1968, President Lyndon Johnson privatized Fannie Mae to get it off the federal budget. It then became a private shareholder-owned company with certain obligations to the public (to make mortgages easier to obtain) in exchange for certain privileges, which included exemption from taxes and oversight, and access to a stupendous line of credit from the U.S. Treasury. Technically, what Fannie Mae does is purchase mortgages from banks where the loans originate. Its sibling, Freddie Mac, was created in 1970 to prevent Fannie Mae from monopolizing the entire secondary mortgage market. Their mortgages are backed by the U.S. government. The existence of these GSEs has diluted, if not eliminated, the discipline inherent to the risky business of mortgage lending, because they are able to purchase such a large percentage of mortgages generated nationwide. The original lenders, knowing they can "flip" the mortgages to the GSEs and be done with them, are far less concerned with the creditworthiness of the borrowers. To the GSEs the borrowers were not even people, merely numbers massed on a video screen. The combined debt possessed by Fannie Mae and Freddie Mac stood at around $3 trillion in 2004, equal to nearly half of the national debt. They are the only Fortune 500 companies that are not obligated to inform the public if they get into financial trouble.

By the time you read this, it is very likely that the housing bubble will have begun to come to grief. With interest rates at rock bottom into the first half of 2004, practically everyone who could have refinanced has now done so. There cannot be another round of re-fi unless interest rates go to zero, which is unlikely to happen and, of course, re-fi doesn't make much sense when interest rates rise, which is what they did in the second half of 2004. In fact, re-fi lending tapered off smartly by late 2004. Housing prices will probably remain inflated for a period of time beyond the end of the re-fi spree because of the end-cycle hangover phenomenon, the persistence of delusional thinking on the part of wishful sellers who refuse to believe that the boom is over and they might have missed out.

In February 2004, Fed Chairman Greenspan made the bizarre suggestion in a public statement that house buyers might consider adjustable-rate mortgages, but the idea seemed insane in a financial climate in which interest rates had nowhere to be adjusted but upward, which would leave many such a house buyer in a terrible predicament of having the mortgage payment go up just when the value of the house had reached its absolute peak and was very likely to fall, as other house owners (especially those with poor credit records, those living marginal lives, those who had lost their jobs since re-fi) lost control of their finances, were forced to sell, or stumbled into default and repossession. Why Greenspan made that suggestion has never been adequately explained. The only possibility is that there was no other way to keep the economy levitated.

The economic wreckage is liable to be impressive. If large numbers of house owners cannot make their mortgage payments, Fannie Mae and Freddie Mac, and by extension the federal government, would be the big losers. The failure of the GSEs would make the S&L fiasco of the 1980s look like a bad night of poker. The failure of the GSEs would pose a far graver situation than the LTCM flameout. It could easily bring on cascading failures that might jeopardize global finance. This time, the American public would feel the pain.

The boom in suburban houses must necessarily be understood as part and parcel of the suburban predicament—the fact that it was part of the greatest misallocation of resources in world history. The entropic aftereffects are likely to be severe. The housing subdivisions, as much as the freeways, the malls, the office parks, and the fast-food huts, represent an infrastructure for daily living that will not be reusable, except perhaps as salvage. I will discuss the destiny of these places in the final chapter.

Reality Bites but Entropy Devours

The global oil production peak will change everything about economic life in the United States (and elsewhere) and especially the value of things believed to be assets, including all paper securities and money. The run-up to the peak, and the squandering of fantastic amounts of energy, has produced a pervasive entropic "pollution" of the entire economic ecosystem. That ecosystem is to some extent a construct of ideas, consensual agreements, and institutions for enacting and regulating those agreements. All that notional infrastructure has supported a fundamental sense of legitimacy—confidence that our shared ideas are sound, that our governments and trading instruments deserve to be trusted, and that we can continue doing what we have been doing—in short, that our ways of behaving have a future.

Where money and its offshoots are concerned, the final entropic consequence of the cheap energy blow-off will be the demise of abstract relations between an asset and what it is supposed to represent. In the Long Emergency, we will be fortunate if enough of the consensus regarding value remains unbroken to have any paper marker at all. The dollar, and any kind of paper associated with it, will be in for a rough ride. The demoralization could easily be worse than that of the Great Depression because we will not be living in "want amidst plenty," as FDR put it, but in hardship amidst scarcity. Our desperate problems with oil and gas will effectively shut down the growth of our industrial economies, and with that our expectations for economic progress, as we have known it. With the consensus about progress shattered, it will be impossible to sustain the illusion that we can get something for nothing.

Mostly we will be preoccupied with concrete local economic realities, and they will bite. Climate change, environmental degradation, falling living standards, and social disorder will be the oil age's gift of entropy to future generations. The transient and ephemeral condition of industrial hypergrowth that the world has known for just over two hundred years will be over. Energy will be at an extreme premium, and human survival skills will be the new capital. What it may be like to live later on in the twenty-first century is the subject of the final chapter.

SEVEN

LIVING IN THE LONG EMERGENCY

I had an odd and illuminating experience a while back, driving from Saratoga ten miles north to the little town of Corinth, New York. The town lies just inside the "blue line" boundary of the state-designated Adirondack "park"—an area actually larger than Yellowstone, but dotted with towns, businesses, factories, a few Wal-Marts, plenty of fast-food establishments, and the usual furnishings of life found in America nowadays. Here in the old Northeast, the land was settled long before the parks movement got going, so the Adirondack Park was an overlay on what already existed.

Corinth (population 2,500) is a paper mill town located on a big bend in the Hudson River upstream of Glens Falls. Above Corinth, the topography gets rugged; the river changes character and becomes increasingly boulder-strewn and riffly. The paper mill closed in 2003 and there is no longer a major employer in the town. No one knows what will become of the town and its inhabitants. For the moment, they seem to get along scrounging a living off the fumes of the cheap-oil economy. They drive long distances on well-maintained state and county roads to low-wage jobs elsewhere—running forklifts in the Target store regional warehouse down in Wilton, being cashiers in the Wal-Mart north of Glens Falls, perhaps frying hamburgers in Saratoga. Or they fix cars, or work on the county highway crew.

On the little two-block Main Street of Corinth, neither the buildings nor the inhabitants look healthy. The buildings show the scrofulous residue of several generations of twentieth-century renovation—exfoliating asphalt shingles from the 1950s, dented aluminum siding from the 1960s, moldy cedar siding from the 1970s (the "environmental" look), and vinyl siding from after that, coated with the inevitable gray-green patina of auto emissions. The shop fronts that are actually occupied—about half the total—contain secondhand stores, hair salons, a pizza joint, and a Chinese takeout. The inhabitants of the town are generally not young. Many are obese, and many of these are cigarette smokers. You see them mostly getting in and out of their cars. No one walks.

I drove up to Corinth along New York State Route 9N, a two-laner much "improved" over the years by the Department of Transportation (DOT) to accommodate 55-mph speeds at every turn. By current American standards it was still a country road, though the roadside itself was chockablock with stuff: houses and businesses, a school, a snowplow garage, gas stations, convenience stores, here and there an old Cozy Cabin-type motel, remnants of the days before theme parks and cheap airfares. The houses were mostly 1950s and 1960s ranch-style buildings, from the bygone era when paper mill workers felt secure enough in their jobs, and were well-paid enough, to build a house—and when they did, it was on cheap rural land along the highway out of town, because there was never any question about their ongoing ability to drive to work. There were a couple of much newer cul-de-sac subdivisions of a dozen houses each along Route 9N, which were built to accommodate those priced out of the Saratoga housing market more recently. These new houses were remarkably cheap-looking and, of course, flat-out ugly in their proportioning and detailing. At intervals I passed derelict farms with broken-back barns and pastures overgrown with poplar scrub. In a nation that had come to subsist on Cheez Doodles and Pepsi-Cola (made possible by giant corn-growing conglomerates such as Archer Daniels Midland and ConAgra) there was no longer any need for local agriculture, even in a region where the farmland was reasonably good and the weather pretty favorable.

If you ventured off Route 9N a hundred yards at any point, you'd be in the woods. But, here's the part that was illuminating to me. As I wended north, ruminating about the Long Emergency and surveying the roadscape, I found myself calculating how each establishment along the road might function in the post-cheap-oil-and-gas world. As I did, I realized that practically none of this would work the way it has been working, if at all. Without cheap gas, how would any of the denizens of these houses range across the vast geographical distances they were accustomed to, driving thirty, fifty, a hundred miles a day to get to a job or fetch groceries? And what would be left for them to get to? Would they have to abandon their roadside entropy bunkers and move into town? If so, how would they sell their devalued houses and what would they move into? If the big-box stores happened to go out of business, and their regional distribution warehouses with them, what on earth would these folks do for a living? How would they heat their houses without liberal supplies of cheap natural gas? Would they strip the Adirondack forests of trees to stay warm? How would their children get back and forth from the sprawling one-story school, and how would they heat it? What would be the purpose of schooling in a post-cheap-oil economy? What kind of careers or vocations would they be training for—surely not public relations or arts administration.

If the folks who lived along this highway put in gardens to make up for the escalating inadequacies of an industrial farming system starved for fossil fuel "inputs," would they be able to feed themselves? Did any vernacular knowledge survive in a populace conditioned to think that food came from the supermarket? Did they know anything about cabbage loopers, powdery mildew, or anthracnose? Would they be able to prevent catastrophic crop loss? How would they defend their crops against deer, rabbits, woodchucks? Would any of them know how to build a garden wall, or even a fence? Where would they get fencing material? Would they have to sit out among the potato hills and the bean rows at night with loaded shotguns? And what would they do for light when they heard something munching out there? Would they know how to keep chicken, sheep, cattle, including breeding and birthing them?

The more I thought about everything I was looking at, the more implausible its future all seemed, and the more fraught with ramifying complications. And of course, the accumulated infrastructure of daily life found along Route 9N between Saratoga and Corinth was minor compared to the vast suburban precincts elsewhere across America, much of it in places where nobody could grow an ear of corn or a potato under any circumstances. What would people in the suburban buzzard flats of Phoenix, Las Vegas, or Los Angeles do when the age of cheap oil and natural gas was over? What will this world be like and what will happen to the people of the United States?

I confess I have had occasion to ruminate on these questions before, but each time I actually do it away from my desk, out in the real world of America where the rubber meets the road, so to speak, where people actually live and work and go about their daily lives, I am confronted by a renewed sense of wonder—and nausea. Sometimes my despondency is overwhelming, and one can well understand why the public hasn't wanted to think about these issues, even in the face of obvious and growing peril. A reasonable person could easily conclude that the way of life we have concocted can't possibly go on much longer, and leave it there. I suspect a great many Americans do exactly that because, so far, by early winter 2004, nothing had changed that much. The massive system seemed to have a momentum of its own that defied occurrences such as a doubling of crude oil prices over the past year. Anyway, life in the United States was so frantic—between the grinding job insecurity, and the war in Iraq, and the horrendous traffic, and orange terror alerts, and child abductions, and the maxed-out credit cards, and the hurricane of the week, and the lack of medical insurance—there was already too much in the here-and-now to worry about.

My role as an author is to think about things that the public is indisposed to dwell on, and to present a framework for understanding a particular set of challenges. What follows, then, is admittedly a personal vision. Some of the ideas I'll present may shock you. Social, political, and economic conditions that Americans assumed had been put behind us for all time may return with a vengeance, especially conditions of social inequality in a world roiled by ferocious competition for declining resources—and in a world with so many weapons available. It bears repeating that just because I say a particular unpleasant thing may happen doesn't mean I want it to happen, or that I endorse its happening. On the whole, I retain confidence in human resilience, courage, ingenuity, and even fairness, and I will spell out comprehensively the positive things that can come out of the difficulties that lie ahead.

First, we have to separate what we wish for from what we're actually doing and what can be done. I believe there is a course of action that is appropriate to what we face, and is actually inevitable, whether we go there voluntarily or have to be dragged kicking and screaming into that future: the comprehensive downscaling, rescaling, downsizing, and relocalizing of all our activities, a radical reorganization of the way we live in the most fundamental particulars. Nothing else will permit us to carry on a semblance of civilized life, most especially not wishing for some mysterious deus ex machina, a.k.a. "them," to deliver a miracle energy source to replace our lost oil and natural gas endowments so that we can continue living in a drive-in utopia. Because human social and economic systems are essentially self-organizing in the face of circumstance, the big questions are how much disorder must we endure as things change, and how hard will we struggle to continue a particular way of life with no future?

The Next Economy

The salient fact about life in the decades ahead is that it will become increasingly and intensely local and smaller in scale. It will do so steadily and by degrees as the amount of available cheap energy decreases and the global contest for it becomes more intense. The scale of all human enterprises will contract with the energy supply. We will be compelled by the circumstances of the Long Emergency to conduct the activities of daily life on a smaller scale, whether we like it or not, and the only intelligent course of action is to prepare for it. The downscaling of America is the single most important task facing the American people. As energy supplies decline, the complexity of human enterprise will also decline in all fields, and the most technologically complex systems will be ones most subject to dysfunction and collapse—including national and state governments. Complex systems based on far-flung resource supply chains and long-range transport will be especially vulnerable. Producing food will become a problem of supreme urgency.

The U.S. economy of the decades to come will center on farming, not high-tech, or "information," or "services," or space travel, or tourism, or finance. All other activities will be secondary to food production, which will require much more human labor. Places that are unsuited for local farming will obviously suffer, and I will discuss this later in the chapter. To put it simply, Americans have been eating oil and natural gas for the past century, at an ever-accelerating pace. Without the massive "inputs" of cheap gasoline and diesel fuel for machines, irrigation, and trucking, or petroleum-based herbicides and pesticides, or fertilizers made out of natural gas, Americans will be compelled to radically reorganize the way food is produced, or starve.

For the past hundred years the trend has been for fewer people to be engaged in farming, and for farming to be organized on an ever more gigantic corporate scale. In that short span of time farming has transitioned from work done by people using knowledge and tools to work done by machines with minimal human presence, almost by remote control. There is a reason that farming is called agri*culture*. The *culture* part stands for the body of knowledge, skill, principles, and methodology acquired over thousands of years. Most of that knowledge has been jettisoned in the rush to turn farms into something like automated factories. In fact, the current system is explicitly called "factory farming" by those who run it. The technology of factory farming promotes the expansion of farms by orders of magnitude above what had been the upper limit for traditional nonindustrial farms. Increasingly farming has changed from being organized on a family and community basis to being corporate and national, even global, with few benefits for the localities where it takes place and with devastating effects on local ecologies and social relations. The diminishing returns of technology in farming have been especially vicious. Few other human activities demand so much respect for natural systems, and the abuse of natural systems has been monumental under the regime of industrial farming. The genetic modification of monoculture crops is only the latest (and possibly the final) technological insult among many previous ones, and comes at the climax of the industrial blowout. Diminishing returns are nature's way of biting back. The "winners" in recent decades have been corporations that could enjoy the economies of scale conferred by gigantism. Their only benefit has been monetary profit. The "losers" can be summarized generally as the future and its inhabitants. They stand to lose not only future wealth, but also their civilization.

All human enterprise can tend toward diminishing returns and unsustainability, but some modes have far more long-term prospects than others and some are socially suicidal, even in the short term. Many civilizations, from the Sumerians to the Maya, have faltered when overinvestments in the scale and complexity of food production produced ruinous diminishing returns. On American farms in the early 1800s, the balance between calories expended and calories produced as food was about even. This occurred as tools reached a high stage of refinement but before machines replaced human labor and traditional knowledge. It implies a distinction between tools and machines, between work done *with* tools and work done *by* machines. Production improved while entropy was kept to a minimum. Under the current industrial farming system it takes sixteen calories of "input" to produce one calorie of grain, and seventy calories of input to produce one calorie of meat.[1] A hundred years ago, just before the introduction of the fossil fuel-based technologies, more than 30 percent of the American population was engaged in farming. Now the figure is 1.6 percent. The issue is not moral, academic, or aesthetic. Rather it's a matter of those ratios being made possible only because cheap oil and automation made up for so much human labor. We did what we did in the twentieth century because we could. Of course, not all farm labor amounts to slavery or serfdom. Depending on how farming is organized, it can result in a very satisfactory way of life and rewarding social relations. Agriculture in the United States was organized very differently in Pennsylvania and South Carolina 150 years ago, and not simply because of climatic differences.

As industrial agriculture reached its climax in the early twenty-first century, the fine-grained, hierarchical complex relations between the soil and the human beings and animals associated with food production have been destroyed or replaced by artificial substitutes. Farmland has in effect been strip-mined for short-term gain. Instead of soil stewardship achieved by acquired knowledge of practices such as crop rotation, manuring, and fallowing, corporate farmers just dump industrial fertilizers and toxins on ground that has been transformed from an ecology of organisms to a sterile growth medium for crop monocultures. Iowa prairie soils 150 years ago had about twelve to sixteen inches of topsoil; now they have only about six to eight inches of topsoil. The loss continues. The "Dust Bowl" of the 1930s was the coincidence of a periodic drought with a decade of zealous overplowing as tractors came broadly into use. The diminishing returns of mechanized plowing were not understood until a catastrophe had been set in motion. The human race had no prior experience with tractors.

A natural lack of rainfall on the Great Plains and in the deserts of California has been compensated for in recent times with heroic amounts

1. Douglas Harper, *Changing Works: Visions of a Lost Agriculture*, Chicago and London: University of Chicago Press, 2001.

of irrigation. In the case of the Great Plains, we've been depleting underground reservoirs (aquifers) of what is essentially fossil water accumulated over the geologic ages and not subject to timely restoration. California's industrial farming system has been made possible by colossal water diversion projects, based on dams (built with fossil fuels) that are silting up and are not likely to continue to function beyond the twenty-first century. The distance people live from the sources of their food has expanded to the extent that today the average Caesar salad travels more than 2,500 miles from the place where the lettuce is grown to the table on which it ends up. Fruit wholesalers in New York City get more apples from Chile than from upstate New York. The few remaining farmers in my part of the state don't even cultivate gardens for their own households. They get their food from the supermarket, like everyone else. Their ecological relationship to the land has been rendered minimal and abstract by technology.

The history of industrialized farming has been remarkably short. Mowing, reaping, and threshing machines powered by animals have barely been on the scene for a century and a half, and engine-driven ones much more recently. The tractor came into common use only eighty years ago. Ditto for electric milking machines and refrigerated bulk storage. In upstate New York, the tractor revolution was not complete until after World War II, that is, within the author's lifetime. Many farmers were still using horses as recently as the 1950s. Yet the loss of knowledge and traditional practice since then has been stupendous. Even if we summon the desire to return to smaller-scale and less oil-dependent ways of farming, the knowledge needed to accomplish it will be hard to reclaim.

Certainly the greatest obstacle to restoring local agriculture, where it is even possible, is the need to reallocate land. In most localities east of the Mississippi, what open land remains near any town or city is considered to have value only for suburban development. It has been so many generations since we have collectively thought about land any other way that our culture will be at a loss to form a new consensus and act swiftly as the times will demand. I daresay we will still be debating zoning issues, such as the right to keep chickens in residential subdivisions, while many Americans starve. Suburbanites may squander their remaining energies in all kinds of futile political efforts to prop up the putative entitlements of suburban living and to preserve the illusion that this way of life can continue.

The result could be years of collective paralysis, indecision, and cognitive dissonance, culminating in social upheaval. In the crisis of the Long Emergency, it will be especially difficult to reallocate or transfer open land already owned to widespread freehold ownership by individual new farmers. The aggressive subdivision of property in the twentieth century has produced an extremely fragmented landscape, even in places that once contained excellent farmland. It will be hard to assemble contiguous parcels into holdings large enough to become farms. There may be more people who wish to resettle on rural land than available land for them, but not enough of them will have the necessary skill to run a farm, not to mention the wealth needed to buy land. One implication is that valuable farmland may tend to remain in the hands of those lucky enough to already be in possession of it, and the less lucky may be enlisted to work on it as hired help. Extreme conditions may lead to the formation of a new peasantry—an exploited class of laboring people tied to the land by contract, custom, or desperate circumstance. That, of course, is essentially feudalism, and while it is not an outcome I savor, it is among the "unthinkable" possible futures that we had better get used to thinking about. Such drastically different social arrangements raise other disturbing issues. What will the role of children be? Will they be part of the workforce? Many of the presumed achievements in social reform of the past century may go out the window with the hardships of a post-cheap-energy world.

I doubt that education would continue to exist as we currently know it. A wild card in a neofeudal scenario might be the potential for violent political upheaval against the propertied classes. Another is the spread of epidemic disease among people already suffering from the stresses associated with plummeting standards of living. Feudalism tends to fail when the supply of surplus labor crashes.

In any case, I'm not optimistic that government could intervene in the reallocation of land for farming. This is exactly the kind of problem against which central government as we know it would prove ineffective, as big government will be subject to the same encumbrances of scale as big agriculture or big business, while the competency of government to redistribute wealth is always questionable—and in the Long Emergency, land will be wealth. If government does attempt to reallocate land on an emergency basis, it might only foment a resistance that would threaten whatever remaining legitimacy it had. Property rights are at the heart of the nation's operating system, so to speak, and to mess with them might be explosive. The Long Emergency will present conditions Americans have never experienced, and the non-rich masses may resort to the kind of desperate action that other historically put-upon people have taken. America is just not that special, nor immune to either the hazards of circumstance or the tendencies of human nature. Revolution might occur, nullifying previous land tenure arrangements, but with central government already disabled it might be limited to some localities and not others. There will surely be a lot of nominally wealthy people left in the nation when the instabilities of post-peak oil kick in, and if their wealth is in land, they may be subject to intimidation, confiscation, or worse.

In the meantime, the agribusiness giants such as Cargill, ConAgra, Archer Daniels Midland (ADM) and others organized like them will go out of business. The food processing industry as we have known it in recent times—the giant national conglomerates that turn ConAgra and ADM's mountains of corn into taco chips and soda pop—will also wither as their financial equations and relations with suppliers fail in the face of oil market disruptions. The supply chains that Americans depend on to magically fill up the supermarkets will be challenged by the coming problems in transport. Food will become much more expensive and far more seasonal. We will be getting fewer apples from Chile in the spring and less lamb from New Zealand in the fall.

If the process of reorganizing agriculture in America were to work favorably—especially in the form of independent freeholds—it could have many benefits. Much farm work will have to be done cooperatively, which would form a basis for a broad infrastructure of social relations, ceremonies, and traditions among neighbors, a kind of "glue" for local communities. Many vocations needed to service local agriculture will open up in craft trades and commerce, which will have additional benefits for community-building. Other occupational niches would be created at the local and regional scales for value-added production—turning milk to dairy products, preserving vegetables and fruits, winemaking, meat processing, breadmaking, and many related activities. In other parts of the world, these economic and social roles were never eliminated in the first place, and the models are there to be seen and emulated. Local distribution networks based on something other than tractor-trailer trucks will have to be reestablished. If we are lucky, these products will be transported over refurbished rail networks or by boat. The retailing of these goods would necessarily have to be done differently, at a finer and small scale, because the giant supermarket chains, the "superstores," and the long-range systems they depend on are unlikely to survive the economic turmoil of the post-peak oil era. Finally, Americans would not be subsisting on the overprocessed junk food that today is a virtual staple for the middle and lower classes, especially the volumes of soda pop that are responsible for the plagues of obesity and diabetes among the poor. Americans will surely be more physically active.

The American scene further into the twenty-first century will have to include more working animals. The horse population in the United States reached its height around 1915 at about 21 million (two years after Henry Ford introduced the assembly-line method of production for his Model T) and declined sharply afterward. The 1920s was a kind of horse holocaust as the automobile and tractor came into broad use and the sudden oversupply of horses sent them by the trainload to rendering plants, like so much scrap. The low point of the U.S. horse population came in the mid-1950s at about 500,000. There are about 7 million horses in the United States today. Of these, 725,000 are used in racing. About 2.5 million are in use on farms—up by about a half-million from the early 1990s. A horse is generally able to begin useful work at four years and it can labor for more than twenty years depending on how well it is cared for. Unlike machines, horses can reproduce themselves. A substantial fraction of production on farms organized around horse power has to be dedicated to growing their feed. Obviously, relations between humans and working animals can range from respectful and loving to careless and cruel, and social norms of decent behavior toward them will have to be reestablished. We are likely, however, to find ourselves living in a world in which this kind of cruelty is more visible. People may be killed by horses, mules, oxen, and bulls, but roughly 50,000 people a year are killed in automobile crashes every year in the United States and there is no public outcry about it.

I don't believe that working animals will replace all the things done by engines all of a sudden, but they are sure to be an increasing presence in our lives, and a time may come when we live with far fewer engines indeed and many more working animals. There is no reason to assume, as we move further away from the oil age, that some miracle replacement for oil will allow a return to industrial agriculture—especially insofar as replacing "soil amendments" made out of oil. You can't make fertilizer or pesticides out of wind power alone. Producing hydrogen by electrolysis from nuclear power and then converting that hydrogen into chemical fertilizers and pesticides would be ridiculously expensive, and even under the best circumstances it would take at least a decade to build a new generation of nuclear power plants dedicated to the task at the necessary scale. We will just have to do farming differently, on a smaller scale, locally, the hard way. An obvious model for this kind of agriculture already exists in the American Amish community, which has stubbornly resisted the blandishments of high-tech through the entire oil-drunk extravaganza of the twentieth century.

The Amish are descended from Anabaptists who emerged from the turmoil of the Reformation. A Dutch Catholic priest named Menno Simons united one branch of Anabaptists into the eponymous Mennonites around 1550. Around 1700 a Swiss bishop named Jacob Amman broke away from the Mennonites and his followers called themselves Amish. Both groups are represented in America. Amish settlement in Pennsylvania began around 1720, under William Penn's "holy experiment" in religious tolerance. Their belief system revolves around the central tenet of being separate from the secular world. This has made them seem more extremely different in custom and manners from other Americans as the twentieth century advanced and a hypersecular consumer society evolved. (A hundred years ago, with a third of the U.S. population on farms, an Amish farming family and a non-Amish farming family might appear superficially similar in ways of working, social organization, and even costume.)

Amish farming practices today are impressively productive and efficient, even carried on without electricity or motor vehicles. Wendell Berry has compared their operations favorably with industrial farming in several books.[2] Can Amish farming practices be separated from the stringencies of Amish religion and social organization? The Long Emergency could provoke a broad renewal of religious observance in America out of misery and desperation, but I do not imagine that any large numbers of ordinary Americans will rush to become Amish. If anything, I expect Americans to turn to the cruder branches of evangelical and Pentecostal Christianity, which will provide simplistic explanations for the dire circumstances in which we find ourselves (and justifications for extreme behavior). I also believe these denominations will seek to reinforce the very hyperindividualist philosophies that evolved with the consumer economy of the twentieth century, and may therefore do a poor job of supporting the kind of cooperative behavior required to restore agricultural communities. Without strong communities based on integral social and economic roles, the revival of small-scale, nonindustrial farming is apt to be haphazard.

I don't know what it will take in the way of spiritual fortification to enable Americans to feed themselves without cheap oil. In a world short of diesel fuel and natural gas, people will have to find other ways to make crops, whatever they believe in. The models are there and the knowledge is there, but it is not in general circulation the way a knowledge of auto mechanics is today. There are plenty of non-Amish people practicing small-scale organic agriculture, and organizations supporting them, such as the Northeast Organic Farming Association (NOFA), which assists small-scale farmers with local marketing, with the preservation of traditional knowledge and technical help, and with political activism to prepare the public for inevitable change. NOFA farmers are often disparaged by an unappreciative public as "boutique farmers," but their activities are vitally important in keeping knowledge alive. There is also a widespread secular subculture of people working commercially in a diverse range of "obsolete" farm-related crafts—harness makers, smiths, farriers, makers of horse-drawn tilling machinery, breeders of draft horses, mules, and oxen—who advertise in periodicals such as *Small Farmer's Journal* and *Mother Earth News*. These craftspeople manage to keep alive skills

2. Wendell Berry, *Amish Economy*, Versailles, KY: Adela Press, 1996; *Home Economics: Fourteen Essays*, San Francisco: North Point Press, 1987; *The Unsettling of America: Culture and Agriculture*, San Francisco: Sierra Club Books, 1977.

that Americans will need desperately when no more trips to the Wal-Mart are possible. The existing literature on small-scale organic farming is vast.

Making a transition out of industrial food production will involve the reestablishment of multiple complex systems on a local basis, including systems of social organization once common in America but surrendered in recent decades. The difficulties of this transition will depend on how rapid the onset of the Long Emergency actually is. I believe that the disorders and instabilities of the post-peak oil singularity will assert themselves rather quickly, long before the world runs out of oil. The quicker they come on, the harsher they will be.

The End of Suburbia

The future is now here for a living arrangement that had no future.

We spent all our wealth acquired in the twentieth century building an infrastructure of daily life that will not work very long into the twenty-first century. It's worth repeating that suburbia is best understood as the greatest misallocation of resources in the history of the world. There really is no way to fully calculate the cost of doing what we did in America, even if you try to tote up only the monetary costs (leaving out the social and environmental ones). Certainly it is somewhere up in the tens of trillions of dollars when one figures in all the roads and highways, all the cars and trucks built since 1905, the far-flung networks of electricity, telephone, and water lines, the scores of thousands of housing subdivisions, a similar number of strip malls, thousands of regional shopping malls, power centers, big-box pods, hamburger and pizza shacks, donut shops, office parks, central schools, and all the other constructed accessories of that life. I have described it at length in other books. The question now is: What will become of it?

Suburbia has a tragic destiny. More than half the U.S. population lives in it. The economy of recent decades is based largely on the building and servicing of it. And the whole system will not operate without liberal and reliable supplies of cheap oil and natural gas. Suburbia is going to lose its value catastrophically as it loses its utility. People who made bad choices and invested the bulk of their life savings in high-priced suburban houses will be in trouble. They will be stuck with houses in unfavorable locations—surrounded by similar dysfunctional artifacts of sprawl—and if they are lucky enough to sell them at all, they will only create an identical set of tragic problems for some greater fool of a buyer. Even fantastic bargains will end up being no bargain. The loss of hallucinated wealth will be stupendous and the disruption of accustomed suburban logistics will be a nightmare for those stuck there. Perhaps a greater question is this: Will the collapse of suburbia as a viable mode of living tear the nation apart, both socially and politically?

The psychology of previous investment implies that enormous political efforts will be undertaken to shore up the illusion that Americans can continue to live in the built environment into which we have sunk our national life savings. Unfortunately, no amount of political legerdemain will keep suburbia running. The ensuing disappointment, hardship, and social turmoil may lead to any of several unappetizing outcomes. Scapegoating is one possibility, blaming various ethnic, racial, or cultural groups for the "problem." This is a very common human behavior in the face of social stress and it could get very ugly, resulting in the persecution or killing of people. It could be undertaken in a political climate of anarchy or under government sponsorship, though I reiterate my point that the post-peak oil predicament is liable to render big government dysfunctional and impotent. I would be more concerned about what might happen on the local and regional levels, and I will discuss that later. Military aggression against other nations is another probability, essentially resource wars, and these are already under way. How long the United States can keep it up is a big question mark.

Exhausting resource wars would only accelerate the plummeting standards of living in the United States and speed the loss of legitimacy of the national government. If large numbers of people cannot unload their suburban McHouses and McMansions, then sooner or later many of these buildings will simply be abandoned, or become the slums of the future. I don't think the transition will be long. Lawlessness may make the continuation of life in the dysfunctional shell of suburbia extra-difficult. The national chain stores will be dead. The supermarkets will not be operating. None of the accustomed large-scale systems we depended on for the goods of daily life will be operating as they did, if at all. It is hard to conceive of any kind of social reorganization that might overcome the practical limitations of the suburban development pattern minus the fuel needed to run it. Many people may try to hang on there, but their lives may be Hobbesian. Where suburbanites might go otherwise is a very good question.

Cities, Towns, and Country

American life in the twenty-first century has the best chance of adjusting to the Long Emergency in a physical pattern of small towns surrounded by productive farmland. I am not optimistic about our big cities—at least not about them remaining big. America's big cities created themselves in tandem with the industrial revolution. They were products of it and servants of it. They were the setting of the first and second acts of the industrial revolution, as suburbia has been the setting for the third act. There was no medieval Kansas City, no Renaissance Minneapolis. The great cities of America, particularly New York and Chicago, became global symbols for the most dynamic and thrilling aspects of everything associated with the "modern," which is to say the cutting edge of advanced techno-industrialism—the glamour of skyscrapers, trains, airplanes, the magic of electricity, movies, telephones, and radio, and all the other miracles of the age. For the non-rich, American cities were always problematic, beginning with the dreary building typologies and continuing to their clunky diagrammatic layouts, their poor street detailing, obeisance to the obnoxious operations of industry, gross commercialization, and finally their abject surrender to the needs of cars. Worst was the sheer overwhelming scale that trapped and oppressed the human spirit. The largest of our cities assumed a scale that had never been seen before in history—as industrialism itself had not been seen before—and this demoralizing hypertrophy produced huge diminishing returns in the quality of life for the industrial masses, especially the workers who crowded the extensive tenement slums. Some of these problems were overcome. The awful sanitation and disease of the nineteenth-century city were vanquished by the great public works of the early twentieth century and the germ theory of modern medicine. Electricity became available to the masses of city dwellers around the same time. Electric streetcars and new subway systems improved life for everyone. The thrill, the charge, the zing of the early twentieth-century hypermetropolis, the business, social, and cultural opportunities, were, for many, compensations for the vicissitudes of scale, the oppression of crowds, and the obliteration of any connection to nature. After 1900, of course, it all depended on cheap oil—and during most of the twentieth century, America was the world's leading producer of it.

But by 1950, the growth of America's big cities was complete and they entered a swift and implacable phase of contraction. Their second act was over. The postwar action was moving to the suburbs. For the next five decades the cities struggled with their losses. Some of them did better than others. New York and Los Angeles retained their dynamism as the financial and media capitals of the two coasts. Los Angeles itself was the prototypical suburban metroplex more than it was a city in any historical sense. While the older big cities contracted, cities based on the post-1950 suburban format grew explosively—Atlanta, Charlotte, Orlando, Houston, Dallas, Phoenix, Las Vegas—but their growth was virtually all suburban. Boston and San Francisco each went into a coma between 1930 and 1980, and then both enjoyed a considerable dynamic revival, and for similar reasons: They became the capitals of the computer industry on their respective coasts, as New York and Los Angeles were the money and media centers. For the rest of America's cities, though, the story in the late twentieth century was much grimmer.

Detroit, which was the world's seventh-richest metropolis in 1950, became, by 1975, a giant suburban donut with a burnt-out hole in the middle. Its middle-class population had decamped to the easily built-upon flat terrain of its hinterlands, leaving the center to a large group of deracinated southern sharecroppers who had migrated to Detroit just in time for the domestic automobile industry to shed tens of thousands of the jobs they migrated there for. By the year 2000, there was hardly any city left where central Detroit had been. Wildflower meadows lay where urban blocks once stood. St. Louis was a similar story, though its decline had actually begun earlier in the twentieth century when it lost out to Chicago as the Midwest's transportation and commodities exchange hub. By 2000, St. Louis was a ghost town surrounded by its suburban donut. Buffalo, New York, was the vaunted "city of the future" in 1900, the "electric metropolis," which, because of the immense generating capacity of Niagara Falls, was expected to become a rival to its older sister at the other end of New York state. By 1980, Buffalo was an economic invalid. The story was similar for Philadelphia, Cleveland, Baltimore, Newark, Pittsburgh, Kansas City, Indianapolis, Cincinnati, Milwaukee. Many attempts were made to rescue these cities, often in the form of skyscraper megaprojects, sports stadiums, performing arts complexes, aquariums, and other grand gestures, but the urban sclerosis just got worse. The third act of the industrial age was equally unkind to the cities occupying the next tier down in scale: Syracuse, Rochester, Worcester, Trenton, Akron, Louisville, Nashville, Des Moines, Chattanooga, and a dozen other once-dynamic small cities all entered the twenty-first century as basket cases surrounded by suburbs that sucked the vitality out of them.

The industrial cities will never again be what they were in the twentieth century. They require too much energy to run and the industrial activities they were designed for are already defunct. Like big corporations, big farms, and big governments, big cities will not be suited to the reduced scale of life in the post-cheap-oil future. Their contraction will accelerate in the Long Emergency. What's more, the superdynamic suburban metroplexes of the past five decades—places like Phoenix, Las Vegas, Houston, Atlanta, Orlando, and so on—will decline even more rapidly and catastrophically than the old industrial big cities once the Long Emergency gets going, because everything in them was designed solely in relation to cars. The biggest cities, New York, Chicago, and Los Angeles, will join Detroit and hemorrhage their populations. They are liable to become dangerously unsanitary and unsafe. Unless the United States ramps up a Project Apollo-style program of nuclear power plant construction, the electric grid is going to be in deep trouble, and there is no question that the North American natural gas supply is already in depletion. The ramifications for big cities are tremendous. As cited earlier, in Chapter 3, the U.S. natural gas pipelines have never been seriously interrupted. The pressure has never dropped below the critical stage at which, for instance, furnaces in large buildings go out. What will happen to the water pipes in a sixty-story residential building in Chicago if the regional natural gas pipeline goes down in February for thirty-six hours? What will happen is that the pipes will burst and every apartment will become uninhabitable. What will happen when the gas pipelines are repressurized and pilot lights don't automatically restart in some buildings? It is a recipe for gas explosions. What will happen in a city full of skyscrapers when the electric grid goes out unpredictably for hours at a time? What will happen to people stuck in the elevators? What will happen to the people down at street level who need to get upstairs to their twenty-ninth-floor office or apartment? What will happen to the elderly? What will happen during a summer heat wave? It's one thing if a blackout occurs in a big city once every fifteen years. It's another thing if it happens every year, or several times a year, or once a month, or twice a week.

The energy disruptions of the Long Emergency are going to remind us that the skyscraper was an experimental building form. These structures operated successfully during the twentieth century, when there was plenty of cheap energy, and after that they became a problem. Economic disruptions will put an end to many of the large-scale enterprises that remain in our cities, and the megastructures that were built for them. There will be no need for headquarters of national companies because, without cheap energy, continental-scale activities will no longer exist. The companies that service the giant corporations in areas like advertising, marketing, and public relations will also wither. Even operations such as the national media and, yes, book publishing as it is currently organized, may not survive in a nation short of energy, crippled in transport, sinking in production and trade, challenged in food production and distribution, and plagued by political crisis.

Urbanist Elizabeth Plater-Zyberk made the cogent observation that cities are where they are because they occupy important sites. Some kind of settlement will continue to exist at Detroit, on the short, strategic stretch of river between two great lakes, and at St Louis, just below where the Missouri River enters the Mississippi, and on Manhattan Island. But they may be nothing like the cities we once knew. In the conditions of the Long Emergency, they may become ungovernable, lawless for a time, perhaps a long time. Disease and unsanitary conditions may haunt them. The biggest cities pose the greatest problems. The idea of physically downscaling Manhattan is close to unthinkable because the island is chockablock with tall buildings that took a century to pile up. They may not be usable in the years to come. But we are going to be a society with far fewer resources in the years ahead and we will not have the means to take them all down and replace them with buildings appropriately scaled to new conditions. Whatever happens in Manhattan or Chicago is going to take a long time to resolve, and may follow a very spooky destiny. Bear in mind that Rome went from a city of more than one million in A.D. 100 to a town of perhaps 15,000 in A.D. 1100. In the eighteenth century—at the dawn of the industrial era—cows were still grazing around the Forum.

Our big cities will no longer have the oil-powered dynamism of their suburbs to lean on for economic sustenance, either. If anything, in places like New York, Chicago, and Detroit, the economic failure of the suburbs will multiply the misery in the cities proper. It may be decades, even centuries before they become civilized or habitable again. New York, in particular, faces daunting problems because of sheer geography. The vast Long Island suburbs appended to the boroughs of Brooklyn and Queens consist of a hundred-mile-long geographic dead end of sprawl-type fabric. If Long Island becomes disorderly, its population may not be able to even physically escape.

In the Long Emergency, the focus of society will have to return to the town or small city and its supporting agricultural hinterland. Those towns and small cities will have to be a lot denser. Most of the towns and small cities of America are in a coma today. The luckier ones, which are generally tourist towns, have had a residue of boutique commerce barely holding the downtown buildings together. Typically, though, downtown buildings in small towns are unoccupied above the ground floors because the landlords will not invest in expensive renovation under strict building codes while new, cheap suburban-style garden apartments pop up on the fringe. The unluckier small towns of our nation—and they are the majority—lie in various stages of dereliction and ruin, their industry gone, their populations aged or idle, the infrastructure rotting. Even solid brick buildings fall apart in a few years when they are not inhabited. Once the roof leaks, all bets are off.

My corner of upstate New York includes both the upper Hudson and the Mohawk river valleys. They were dotted by what used to be, until about 1960, a series of vibrant manufacturing towns—places like Hudson Falls, Fort Edward, Glens Falls, Amsterdam, Johnstown, as well as small cities such as Troy, Schenectady, Utica, and Syracuse. Little Corinth, which I described at the start of this chapter, is one of them. Today these places stand gutted, half-vacant, idle, hopeless. Paradoxically, they may be the kinds of places that have the strongest chance of surviving the challenges of the Long Emergency. It won't be easy. But they have the potential of coming back to life at the scale that new economic realities will require. And as populations decamp the suburbs and the metroplexes, these are places that might attract them.

These towns are located on rivers and, in the case of the Hudson Valley, with its rugged topography, they have many potential sites for local hydropower generation. They are surrounded by good farmland that remains substantially intact. The downtowns and old residential neighborhoods are laid out along the lines of pre-cheap oil urbanism, compact, dense, and walkable, and most of them are intact, though the buildings are badly run-down. They are not burdened excessively by skyscrapers and other megastructures. They can be much more easily fixed and infilled than places built from the suburban template. The increment of redevelopment in the Long Emergency is going to be small, one building lot at a time—the age of the megaproject, along with megafinance, will be over. But at least the projects will be doable, and the doing can be done by the people who live there, not by corporate developers from elsewhere looking only for short-end money. In the meantime, if local agriculture can be reorganized and revived, the new inhabitants of smaller cities and towns have reasonable prospects of feeding themselves, and finding useful vocations geared to local activities. They also have a better chance of maintaining civil order than either the big cities or the sprawling suburbs, though population shifting itself is liable to entail desperate competition for resources and perhaps even conflict.

The people who lived in the small industrial towns of America one hundred years ago never would have believed what became of these places at the turn of the new millennium. The desolation and loss would have been inconceivable to a people who had done such a fine job of building communities. Even from the closer vantage point of 1950, the destruction has been incredible—as though World War II had been fought in Schenectady, New York, rather than Bastogne. Tragically, the destruction was all the result of economic suicide, of bad decisions made for bad reasons, by people grown too complacent and greedy to care much about the future. In the Long Emergency, there will be a tremendous cultural reaction to the epic carelessness of those generations responsible. Their legacy will be looked on not with nostalgia, but with contempt and amazement.

The psychosocial infrastructure of our communities is going to also change a great deal. One of the basic confusions inherent to the suburban experiment was the idea that people could live an urban life in the rural setting. The Long Emergency will revise that. People who live in rural areas, or resettle there, must prepare to lead rural lives and follow rural vocations. That means food production, farming. Those who live in towns, even small towns, will work at activities appropriate to the town: trade, education, medicine, and so forth. The distinction between town and country will be much less ambiguous. There will be no more twenty-six-mile drives to the Super Wal-Mart or the vet.

Much of America east of the Mississippi is full of towns like the ones here in upstate New York where I live. They are standing there, waiting to be reused, with much of their original equipment intact. The lucky suburbanites will be the ones with the forethought to trade in their suburban McHouses for property in the towns and small cities, and prepare for a vocational life doing something useful and practical on the small scale, whether it is publishing a newsletter, being a paramedic, or fixing bicycles. I will make a distinction in a section ahead between the destiny of the Old South and the northern states of what I call the Old Union.

Commerce in the Long Emergency

Just as farming evolved to reach gigantic scales of operation and adopted corporate modes of organization the past half century due to the economics of cheap oil, so did everyday commerce evolve to a colossal scale under monopolistic global corporate enterprises that ruthlessly destroyed complex local and regional networks of economic interdependence. I have already described how the process occurred and the character of the national retail chains involved. The Long Emergency will put all of them out of business. Wal-Mart, Kmart, Target, Home Depot are all going to wither and die. They were strictly manifestations of the cheap oil final blowout and they will not survive beyond it under the conditions of the

Long Emergency. Wal-Mart will not be able to profitably run its "warehouse on wheels" when the price of oil fluctuates chronically (always along an upward trend line), and supplies become less than completely reliable. Wal-Mart and the others will not be able to maintain relationships with suppliers twelve thousand miles away in China when we are locked in desperate competition with that nation over oil supplies, or when the Asian shipping lanes are effectively shut down by anarchy on the high seas. Even if Wal-Mart could get ultracheap manufactured goods from some other place in the world, it may have no one to sell to when the American middle class becomes an impoverished former middle class, and the current consumer credit structure has cratered.

The idea of a consumer culture itself is going to die with the national chain stores. We will never again experience the explosion of products, choices, and nonstop marketing that characterized the late twentieth century. The public may look back on the big-box shopping era with deep and mournful nostalgia, but we are apt to discover that happiness is still possible without the extraordinary advertising-driven compulsive materialism of recent decades. We will still have commerce. We will have trade. There will be shopping. We will have some kind of medium of exchange. But we are not going to live in a perpetual blue-light special sale of cornucopian wretched excess. We will not be consumed by our consumption. Nor will our children or grandchildren.

We are going to have to rebuild from the bottom up those complex webs of local economic interdependence that the national chain store movement has destroyed. It is a tremendous and daunting task. The transition out of corporatist hyper-retailing will be very painful. If political friction with China were to occur rapidly—say, in a crisis over Taiwan—we might even endure a period of chaos in the everyday markets. We became addicted to ultracheap household goods as surely as we became addicted to imported oil. Many common products we depend on aren't manufactured in the United States anymore—all sorts of things from bicycle tires to dish towels. We are not going to restart factories for these things overnight. In fact, because we will be living in a far lower-energy society, our basic capacity for manufacturing anything will necessarily be much lower. Many of our pre-1945 factories have been either demolished or stand in a state of hopeless dereliction, and I would argue that we will not be able to do any kind of manufacturing based on the scale of a 1950 factory model anyway. What manufacturing we do manage to reestablish may be what used to be called cottage industry, based more on craft skills than the assembly line method of production. We may have to do without a great many of the products we became accustomed to during the last half-century. In addition to these problems of basic commercial and manufacturing infrastructure is the case that so many of our common products, tools, appliances, paints, chemicals, building materials, pharmaceuticals, and especially plastics have been made directly out of oil or natural gas.

We should be able to supply the basic necessities of life for ourselves. But even the term "basic necessities" is elusive. What were basic necessities for Thomas Jefferson and his contemporaries would seem extremely meager to many of us. Imagine life without insect repellent, air conditioning, and flush toilets. It is hard to predict what the level of comfort may be in the Long Emergency, but I think we can count on something a bit more advanced than the level of the eighteenth century. Even if we can't get all the tools and the products we currently enjoy, we will retain a lot of basic knowledge that the people of Jefferson's day just didn't have. For instance, we will still understand that infections and many diseases are caused by microorganisms, not bad air, phases of the moon, or evil spells, and that knowledge alone confers powerful advantages in daily living.

Another thing we can count on is the tremendous inventory of recyclable material and manufactured goods already in existence in the United States. The closets, basements, attics, and garages of America overflow with so much stuff that outside-the-home self-storage is a booming business. One of the major areas of useful work in the Long Emergency will be the repair and resale of goods. Human beings have a genius for sorting useful materials. Americans have been fantastically profligate in throwing out goods and appliances that are only slightly broken. We don't call a repair person to come over and fix our two-hundred-dollar color TV sets. We put them out on the curb with a sign that says "FREE" taped to the screen and drive over to the Best Buy store to get a new one. We throw computer printers in the garbage when they malfunction. This will change radically. In the Long Emergency, there will be far fewer new things and less stigma attached to buying things that are not brand new.

The conditions of the Long Emergency will militate against corporate organization as we have known it, that is against commercial enterprises scaled to operate virtually like sovereign states run by oligarchies. Whether their demise is a good or a bad thing remains to be seen. Large-scale corporate enterprise has brought humankind much material comfort in two centuries, but at the price of fantastic unintended consequences (externalized costs) ranging from the destruction of local communities to climate change. Large-scale corporations will be vulnerable to the collapse of capital formation markets that must accompany the end of the cheap-oil fiesta. Corporate enterprise can certainly be reorganized on the small, local community scale, but it will not be the same as General Motors. Corporate enterprise in the Long Emergency may revert to being more public in nature and far less sovereign in power. There may be one exception: The most visible kind of corporate organization that might survive the Long Emergency may be the church. Whether Catholic or Pentecostal or something new we haven't seen yet, the church won't have to rely on oil supplies. Organized religion doesn't have to traffic in awkward material products, only in beliefs, and it can operate at many scales simultaneously. Because American culture is constitutionally allergic to religious governance, we may have problems if churches are the only large organizations left standing—that is, assuming we still have the same constitution.

Whatever the content of our everyday commerce consists of in the way of merchandise, either new or recycled, it will require an operating system and an infrastructure. The model is obvious. Since we will need to live an intensively local existence, we will have to reestablish those local webs of economic relations and occupations that existed all over America until the last several decades of the twentieth century, meaning local and regional distribution networks, which would include retailers, wholesalers, jobbers, warehousers—all the intermediaries who were eliminated by the national chain stores. The propaganda of the chains always claimed that the public—degradingly referred to as "consumers"—benefited from the elimination of these intermediaries because the chain stores could "sell direct" and charge wholesale prices for their goods. The bad end of the bargain, of course, was that so many Americans had been engaged in precisely those eliminated jobs that a substantial portion of the middle class was left without livelihoods—exactly those business-people who had the greatest stake in maintaining local communities. As these positions are restored in the vacuum left by the fall of the national chain companies, many of the civic caretaking roles in our communities will be restored with them.

If locally owned retailing is restored, chances are that the people conducting it will take care of the buildings they occupy. They will know the people they do business with and they will be very careful to maintain respectful, friendly relations with them. They will employ people in the community and there will be more social pressure to treat them responsibly, especially when the old government safety net is no longer in place.

What We Live In

It is hard to say how much of the existing late-twentieth-century building infrastructure might be adaptively reused in these reestablished local economic networks. There is a lot of it. More than 80 percent of everything ever built in America was built after World War II, and most of it was designed solely to be used in connection with cars. There will probably be three considerations: (1) how walkable it is—in many places, especially the metroplexes of the Sunbelt, the strictures of zoning have left retail buildings extremely isolated from both the residential neighborhoods and the original town centers; what had seemed minutes away in an air-conditioned car may feel more like the Bataan Death March on an August afternoon when one must get there on foot; (2) whether the buildings can be heated, or need to be heated—given the depleted state of the North American natural gas supply, it's unlikely that the abandoned big-box buildings will be heatable, which obviously would be more of a problem in the northern tier of the country; an empty Kmart in Biloxi, Mississippi, can, and probably will, become anything from an infirmary to a Pentecostal roller rink, but in Wisconsin it's likely to be a different story; (3) whether the roofs can be kept in repair—unfortunately, most of the commercial structures built in America after World War II have flat roofs. They are penetrated by all kinds of vents and mechanicals, and generally these have been sealed using oil-derived materials over synthetic rubber flashing.

Even with regular maintenance, their design life is probably twenty years at most. Once water gets in under the roof, deterioration proceeds very quickly. Electric wiring is especially susceptible. It is hard to say whether the snow and ice in the colder states would be more damaging than the extreme ultraviolet light and torrential rains of hot places such as Florida or Texas. We should probably conclude that the abandoned big-box structures will not last more than one generation under any circumstances. Pretty much the same thing can be said about malls, strip malls, and chain restaurant buildings. Eventually they will be the salvage yards and mines of the future.

Our communities therefore will have to reorganize physically as well as socially and economically. The existing small cities and towns of America offer the best opportunities for that, as they already contain the appropriate building types deployed in an appropriate walkable street-and-block template. Many of the building codes created over recent decades will probably have to be ignored or abandoned as communities seek to rebuild within the context of a much more austere economy. These building codes were extremely restrictive, especially the recent handicapped-access laws, which required elevators in virtually every building over one story that was not a private house. Because commercial builders of this period generally chose the path of least resistance, this tended to result in nothing but one-story buildings. State fire codes also mandated extremely onerous requirements for multiple stairwells in buildings over one story—often to such a ridiculous degree that they would take up most of the internal space of a proposed building, thus defeating the purpose. This was particularly true where the renovation of existing older buildings was concerned, including Main Street buildings in towns all over America. By the time a renovator theoretically finished putting in elevators, handicapped ramps, and multiple egresses, there would hardly be any room left for apartments or offices in the building, or they would be awkward to organize around all the internal infrastructure. These excessive regulations were devised not just for safety and fairness, but for reasons of legal liability. They represent, more than anything, a fear of lawyers.

In the Long Emergency we will not be able to afford this over-regulation. We will no longer be living in a strictly horizontal environment defined by cars, parking lots, and one-story buildings deployed arbitrarily all over the landscape. Our daily environments will have to be much more defined by walking distances. Our towns will have to be much more compact. They will be more vertical—within limits. We will have to get a much greater percentage of our two- to five-story buildings back in full service, and we will have to build new ones to fill in the spaces of the old parking lots. As I've already suggested, the skyscraper will be an anachronism. Buildings above seven stories may be out of the question for practical reasons. If we abandon the building codes of the twentieth century, our buildings may become less safe. Winter heating is also likely to be a big problem, not only in terms of available fuel, but also because we will have less central heating and a greater emphasis on individual room heaters, which tend to be more dangerous as a rule. This may have additional implications for our plumbing needs. Modern plumbing went hand-in-hand with central heating. If you can't heat an entire building, the pipes are likely to freeze somewhere, and if they freeze at even one point, the whole building has a problem.

Many of the modular construction materials used by commercial builders—the plastic artificial stuccos, fabricated epoxy panels, and so on—will no longer be available. We will probably have to return to traditional masonry and wood construction. Reinforced concrete may not even be possible if there is a shortage of steel "rebar." Some materials may come from the disassembled strip malls, big-box stores, and other obsolete structures of the twentieth century. As I've already suggested, the increment of redevelopment will be very small compared to the scale we're accustomed to. One new building on a single lot may be a big deal in the Long Emergency. There may be relatively little building activity per se, and the construction "industry" as we have known it almost certainly will not survive—certainly not the so-called "production housing" sector.

Local communities of the kind I have described, where businesses are owned by the people who work in them, and where social and economic roles form a rich matrix of interrelationships, are generally good at creating local institutions to care for those who are weak, disabled, or old—who are their neighbors. America, in fact, was full of such local institutions in an earlier period. Much as they acquired a kind of folkloric stigma, places like poorhouses and county farms of the nineteenth century were arguably far more humane than the conditions that the weak and indigent are subject to today. These institutions were often organized to be nearly self-sufficient in food, with the inmates or patients doing the garden work and household chores themselves. Norms of social behavior were enforced. Vested authority was taken seriously. Throughout the twentieth century, these caretaking functions were assumed by the states and operated at a gigantic scale that was inherently inhumane. The infamous colossal state mental hospitals of New York typically housed more than five thousand patients (or inmates—whatever term you choose will acquire a stigma). Patients were treated like ciphers and were lost in the back wards. By the 1970s, these institutions came to be regarded as failures. The state closed them up. Unfortunately they were not replaced with small-scale local institutions. Instead, the patients/inmates were released to the streets, where in our time they became, simply, the homeless. We will almost certainly have to recreate caretaking institutions at the local level, in bricks and mortar, and they could be much better places than anything we have seen in a long time. They will also, of course, create caretaking jobs in our communities.

The character of our communities and the activities that go on in them will be greatly determined by how we are able to get around—how people and things might be transported in the post-cheap oil economy. We can be confident that the current set of arrangements will not endure. Everyday life deeper in the twenty-first century will be as starkly different to people living today as the America of 1955 was different for someone who was a child before World War I. The automobile age as we have known it will be over.

Transportation in the Long Emergency

The twenty-first century will be much more about staying put than about going to other places. This idea is astonishing to many Americans today, who have been in motion on wheels more or less continuously for their whole lives, whizzing from driveway to parking lot to curb-cut to carwash to big-box store and home again, often several times a day, year in and year out. Not many years from now, the automobile will be a much-diminished presence in our lives. I believe cars will still exist, but in far smaller numbers. They may run on various things, including whatever non-cheap gasoline or diesel is still available. As discussed in Chapter Three, replacing the current internal combustion car fleet with hydrogen-powered fuel-cell cars is unlikely to happen under the current laws of thermodynamics. Cars using regular electric motors are a better bet, though the battery problem limits their range, and to some extent their existence in any numbers is predicated on a renewed nuclear power effort. That in itself may be impossible to accomplish in a nation with an impotent central government.

There are additional problems with keeping the car-and-highway system, as we have known it, going through the Long Emergency. One is political. As the middle class becomes increasingly distressed economically, and oil supplies become more expensive and possibly irregular, using a car will become more and more a luxury. If car use becomes something only for the elite, it is apt to excite the resentment of those whose driving opportunities have been foreclosed by economic misfortune. That resentment might become extreme. Cars might be vandalized. Drivers might be subject to physical abuse.

The American motoring system has been able to work because everyone from the lowliest burger flipper to the richest CEO could participate. The costs of running it were democratically borne by everybody. Will economically hard-pressed citizens who can't keep a car going tolerate their taxes going to maintain highways that only the elite can afford to drive on? If tax revenues decline broadly along with incomes, and don't cover the cost of repairing highways, where will the money come from? How will we fund all those state DOT projects?

The interstate highway system is more delicate than the public realizes. If the "level of service" (as traffic engineers call it) is not maintained at the highest degree, problems multiply and escalate quickly. Sixty-four percent of all freight in America is moved by truck. A pavement that is only slightly broken up quickly deteriorates under the continuous pounding of tractor-trailer trucks. The system does not tolerate partial failure. The interstates are either in excellent condition, or they quickly fall apart. Once they do fall apart, serious economic consequences ensue. Trucking becomes a losing proposition. Broken pavements mean that heavy trucks are subject to breaking their axles—about the most dangerous thing that can happen to a truck at full speed—as well as to aggravated general wear and tear.

Of course the trucking industry faces its own economic hardships, apart from the condition of the roads, every time the price of oil goes up even a small amount. Many of today's truckers are independents working under contract to large corporations. The catch is they must bear all the costs of doing business, including fuel, insurance, and making payments on their expensive trucks. The global production peak will lead to such radical price and supply instability that many truckers would simply go out of business. They will not be able to economically rationalize the costs of hauling. Any business that depends on transporting massive volumes of merchandise on reliable "just-in-time" schedules, cheaply, in order to justify razor-thin profit margins, would also be in trouble. The vaunted "efficiencies" of Wal-Mart's "warehouse on wheels" would vanish. American commercial transport would quickly tumble to third-world levels of inefficiency and expense.

I believe that the interstate highway system will reach a point of becoming unfixable and unmaintainable not far into the twenty-first century. The resources will not be there to keep up the level of service at the minimum necessary to prevent cascading failure. I think we will be shocked by how rapidly its deterioration proceeds. It will happen at the same time that the economics of mass car ownership become untenable, and the two failures will be mutually reinforcing. The decline of our highways will be worse in some regions than in others, probably worst in the largest suburban metroplexes, where repair will not be able to keep pace even with decreased levels of motoring. Not even counting the interstate highway system, there are more than 3.8 million miles of paved roads in the United States, all of it funded politically. Today, while the United States is still relatively flush, the federal DOT considers 18 percent of federal highways to be in "poor" or "mediocre" condition, whereas 29 percent of the thousands of bridges are "structurally deficient" or "functionally obsolete."[3] In the Long Emergency, the highways will still be there. They may continue to be usable at low speeds by cars and other vehicles for a long time after the initial process of failure

3. P. J. O'Rourke, "Pork with a Point: The Highway Bill, a Translation," *The Atlantic*, December 2004.

begins. But we will no longer buzz around on them unthinkingly at 70 mph the way we did in the late twentieth century.

Those in charge of the highway system at all levels have not been able to plan for these epochal discontinuities because doing so would imply the demise of the system as we know it. Instead, they have applied their theoretical thinking on ultra-high-tech management of infinitely increasing levels of traffic. The so-called "intelligent highway system" (IHS), for instance, was the chief preoccupation of highway engineers oblivious to the approaching peak oil catastrophe and entranced with the computer advances. Their fatuous expectations of meeting ever-increasing traffic loads with ever more sophisticated onboard car computers keyed to computers embedded in the roadways couldn't have been more detached from reality. Even on their own terms the schemes were laughable. They assumed that every car on the road would have an onboard computer. One had to wonder: What about the drivers only pretending to have onboard computers, just as so many drivers today only pretend to have licenses and insurance? Proposals such as IHS demonstrate how overinvestment in technological complexity can continue far beyond the appearance of obvious diminishing returns. America's paved highway network as a whole was an amazing achievement. There were only 141 miles of paved roads outside the cities in 1904. Modern highway building had barely begun by 1920. The entire project was completed in my parents' lifetimes. It was assumed to be a permanent installation. It will prove to have lasted barely a hundred years.

An obvious answer to the decline of the car-and-highway system would be to revive the American railroad network—once the envy of the world, now a world-class embarrassment. What's not so obvious is whether this is even possible. For one thing, the suburban development pattern based on the car has evolved in a way that is exactly the opposite of the centralizing nodal pattern of towns and cities that tends to self-organize around railroad service. Assuming that can be overcome—which is assuming a lot—there are many sheer logistical problems. Merely rebuilding the steel rail lines, or repairing existing ones, would be a formidable task in a nation challenged by depleting energy resources. Manufacturing steel rails requires heroic quantities of energy. We could conceivably direct our coal supply to this end, especially in the early years of the Long Emergency, but our steel manufacturing capacity has been surrendered almost completely to other nations. We'd have to restart a domestic steel industry, and time to begin planning for this is running short.

Most existing American train engines today run on diesel fuel, and it might be cost-effective to keep running them, even if the price of crude oil soars, as trains are exceedingly efficient for moving people and freight. Scarcity is another issue. Diesel fuel is just another product of the oil refineries, after all, and would be subject to the same market instabilities as gasoline supplies. As we know from history, railroad engines can also run on coal. They would produce more carbon dioxide and particulate pollution than engines run on diesel, but the number of engines running at any one time would be minuscule compared, say, to the number of cars on the road at any moment in 2004. Electric motors have fewer moving parts than diesel or steam engines and are less subject to breakdown or explosion than coal-fired steam engines. Electric motors would not create air pollution directly, but, of course, nuclear power raises an entirely separate range of associated problems. For instance, why not just devote nuclear power to electric cars? With cars the battery/range-of-travel issues still pertain. With a railroad, the electricity supply runs continuously along the track, overhead or on a third rail, so the range of an electric engine is theoretically infinite. Before World War II, U.S. railroad electrification was about on par with other industrialized countries. After the war, not only did U.S. electrification come to a standstill, but it also started to decline and be dismantled. In Europe, the Soviet Union, and Japan electrification resumed with renewed vigor. The result was that the United States wound up by the end of the twentieth century with no major electrification of freight lines. Only Amtrak, the government-subsidized passenger service, had substantial electrified passenger service between Boston and Washington, D.C. In Europe today, 40 percent of the railroads are electric.

If we made the choice to rebuild the U.S. railroad system, in an austere economy the infrastructure would be much less onerous to maintain than our vast highway system. The trains could be powered with electricity produced by nuclear power plants. Though I believe nuclear power would soften the post-cheap oil crash, I would not bet on the likelihood of America getting a new generation of nuclear plants ramped up in time. Therefore whatever railroad system we manage to cobble together would not be electrified. In addition to the administrative problems in organizing a nuclear program (arguably greater than those faced by the national railroad project) is the fact that an underlying oil economy would be required to build and run a comprehensive system of nuclear reactors.

Similarly problematic is whether the project of restoring America's railroads could be organized at a scale equal to the gigantic task. In the Long Emergency, all large-scale enterprises will have trouble operating in virtually every sphere of activity. Rebuilding the railroads is a project perhaps comparable to building the interstate highway system, or nearly so. Could that be done today from scratch? Or was it something that could have been brought off only in a particular moment of history—by a hyperaffluent nation awash in cheap oil led by a generation of victorious World War II veterans schooled in the heroics of war production? We certainly can't return to the special economic conditions of nineteenth-century America that made the initial building of railroads possible—namely, huge amounts of open land or sparsely settled land, coupled with federal right-of-way land grants, allowing railroad companies to profit greatly in real estate on the side.

One answer to this dilemma might be that the U.S. railroads in their glory days never ran as a single monolithic system. Many dozens of regional lines were stitched together, employing standards of gauge, switching, and operating protocols that made the combined organisms seem to function as a single system. It might be argued that the attempt to run freight and passenger rail systems under the near-monolithic administration of Conrail and Amtrak in recent decades has been a management fiasco precisely for reasons of excessive scale.

If the American railroads are to revive at all under the conditions of the Long Emergency, they will probably do so in a piecemeal way as a stitched-together patchwork of regional lines. Some will operate better than others. Their operation could be hampered by social turbulence and political unrest. One final thing worth noting on the subject of rail: From 1890 to about 1920, American localities managed to construct hundreds of local and interurban streetcar lines that added up to a magnificent national system (independent of the national heavy rail system). Except for two twenty-mile gaps in New York state, one could ride the trolley lines from New England clear out to Wisconsin. The story of the conspiracy by General Motors and other companies to destroy the U.S. interurban system is well documented. The salient point, however, is how rapidly the system was created in the first place, and how marvelously well it served the public in the period before the automobile became established. Light rail of some kind—which does not require elaborate roadbeds—may become the basis for regional transportation systems of the future. The lines can be laid along the existing right-of-ways of any of our roads, from the interstates to the streets of our towns, and they can usefully transport people and freight. If energy conditions become really dire, they can use draft animals to convey the cars more efficiently than wagons on roads with broken pavements.

U.S. waterways are probably the most forgotten and neglected elements in our national transportation system. Who, living in the year 1920, would ever have believed that the entire waterfront of Manhattan Island would be devoid of commercial docks at the end of the twentieth century? Or that Boston's harbor would host only a few tour boats? Our current wet dreams about turning so many waterfronts into parks is a good indication of how shortsighted even well-intentioned civic leaders have been in our time. If we want to conduct trade further along in the twenty-first century, whatever our commerce consists of, it will have to rely much more on water transport. It will be slow, but it will be doable, and where inland waterways are concerned, dependable. It would have to be integrated with whatever rail systems we can put back in service. Parts of the American West lacking navigable waterways will obviously not benefit.

Where I live the extensive Champlain and Erie canal systems are still intact. In fact, they continue to be maintained impeccably despite the very low volumes of traffic. These days, they are used mainly by a few recreational boaters bringing their yachts up north for the summer from Florida. Once in a while an oil tanker passes through. The locks could be operated by locally generated power and therefore not have to depend on a regional grid. Most locks of the Champlain Canal still have generator houses standing, though the equipment inside is long gone. The power would have to run only when the lock gates needed to open and close.

It's hard to be optimistic about the future of commercial airlines. Fuel represents about a quarter of their operating costs, but even at 2004 price levels, with fuel supplies still very dependable, the commercial airlines were struggling to stay in business. Like so many other enterprises that operate on a gigantic scale, they depend on a high-volume business to make their economies of scale work. But if the economics of scale change the way they are likely to, they are toast. For instance, how would the airlines cope when the broad middle class can no longer afford casual flights to Orlando and Las Vegas, or just to visit Grandma in Louisville? When that happens, of course, the business-class fares airlines depend on will fall off as corporations tighten their belts. Since the Long Emergency will be open-ended timewise, the airlines will enter a fatal spiral from declining volumes of business. As the Long Emergency progresses, air travel will be increasingly the province of the wealthy, flying private aircraft, and if the social turmoil is bad enough, many of them will be winging out of the United States with no immediate plans to return. Ultimately, commercial aviation as we have enjoyed it may become altogether a thing of the past. The Wright brothers managed the first sustained flight not because their vehicle design was so brilliant—lots of people understood Bernoulli's principle—but because they applied a suitably powerful gasoline engine to the problem. Jetliners manage pretty much the same thing on the grand scale but with turbines powered by aviation fuel, which is essentially kerosene. Kerosene has been fairly cheap in recent decades. It is just another product of petroleum distillation like gasoline and diesel. Commercial jets burn up fantastic quantities of it. If the price goes way up while the number of passengers goes substantially down, the airlines will have no economic reason to continue operating.

Any way one might imagine it, the transportation picture in the mid-twenty-first century will be very different from the fiesta of mobility we have enjoyed for the past fifty years. It will be characterized by austerity and a return to smaller scales of operation in virtually every aspect of travel and transport. It will compel us to make the most of our immediate environments.

Education

It's hard to imagine a more purposeless activity than American-style high school in our time. I doubt that the public questions its basic premises or mode of operation any more than the public questions the economy of suburban sprawl. But high school in our time amounts to little more than day care for virtual adults in which some learning might incidentally take place, much of it of dubious value. There are any number of rationales that might explain it—for instance, that all young people must be prepared for college, because the knowledge economy demands a highly educated workforce. This is probably fallacious because, in fact, most nonmanual-labor jobs (including many decent and useful ones) do not require anything more than the ability to write a coherent paragraph or perform a few rudimentary operations of arithmetic—which is asking a lot, by the way, as quite a few graduates of American colleges cannot do either very well. Another rationale is that secondary education is a way of usefully occupying young people who would otherwise clutter up the job market—and most states have laws that prevent people under age sixteen from working full-time jobs. Mostly, though, we don't think much about the ethos behind schooling today. Like so many other everyday activities in America today—commuting, TV watching, lawn mowing—we accept the current model of education as absolutely normative and inevitable. It has worked a certain way for the better part of a century and nothing has really provoked us to rethink it, only to make it ever larger and more democratically inclusive.

Yet the failure of schooling in America is manifest. Our inner-city schools are in a nearly complete state of entropic decay due to the effects of our overall disinvestment in cities—the school buildings themselves are crumbling, while books and supplies are beyond the point of critical shortage—and to an array of social conditions ranging from the disintegration of families to the absence of standards of normative behavior. Whether these might all be lumped together as the consequences of poverty is debatable, in my opinion, but the effects are not debatable. These schools are not producing even minimally literate citizens with adequate social skills.

Suburban schools may be in better condition physically, with more abundant supplies, but that, too, is changing. Gigantic alienating suburban schools are producing so much anxiety and depression that multiple slayings have occurred at regular intervals in them in recent years. All school in America, whether inner city or suburban, is still based on the obsolete model of a 1911 factory, with several thousand workers in attendance whose routines have to be regimented. Its chief mission is custodial, despite the putative reforms of recent decades. It conditions children and young adults to spend the bulk of their day in one place. It uniformly infantilizes young adults, including even many of those who succeed on its terms. The rest merely suffer in it, while losing the opportunity to learn manual skills that would make them useful and productive members of a community. The system will not endure much longer in this form.

The huge centralized suburban schools that look like medium-security prisons with their fleets of yellow buses will rapidly become obsolete when the first serious oil market disruptions of the Long Emergency occur. The inner-city schools will be too broken to fix. School will have to be reorganized on a local basis, at a much smaller scale, in smaller buildings. Children will have to live closer to the schools they attend.

In the Long Emergency schooling will be required for fewer years, and children may have to work part of the day or part of the year. Because everything will be local, the ability to support education will depend on local economic conditions and the level of social stability, and there will be broad variation. Some localities may become so distressed that public school will cease to exist. The more fortunate localities will be those where small-scale agriculture is possible, but more intensive local agriculture by nonindustrial methods implies a much different division of labor, and older children may have to assume more responsibility and grow up faster. The romanticization of childhood may prove to have been one of the luxuries of the cheap-oil age. Basic schooling, in the formal sense, might not go beyond the equivalent of today's eighth grade. Sorting of children into vocational or academic tracks will probably be based on self-evident social and economic status rather than any formal administrative system. Only a tiny minority of young people will be able to enjoy a college education. Vocational training is much more likely to occur in the context of a workplace rather than the school, as in the apprentice system.

However many grades education entails in the future, children and teachers would benefit tremendously from being in physically smaller institutions, in smaller classes, where all will at least have the chance to know one another. Many of the problems of education today are the unintended consequences of consolidated administration—the attempt by school districts to save money by operating fewer but bigger buildings, running fewer but larger bus fleets, and employing fewer nonteaching managers. This was made possible by a chain of connected circumstances. The suburban development pattern erased the essential quality of locality per se. Typically, by the 1960s, suburban children couldn't walk anywhere, including to school, so wherever they did go to school a bus was required to get them there. Cheap oil made the school bus fleet a normative part of the system. (Think of a school bus fleet as a public transit system that runs only twice a day for people under eighteen and you may grasp the basic profligacy of the system.) Once that was established, school districts gathered their pupils from ever-larger geographic population "sheds" and bused them to ever more gigantic consolidated "facilities." The economies of scale they strove to enjoy were typical of any large enterprise during the cheap-oil age. The effect of all this on the students, though, was always secondary to the administrative benefits, and the purpose of school somehow got lost, so that, paradoxically, even the richest suburban high schools with Olympic swimming pools, food courts, and hectares of playing fields produced alienated students dogged by anomie, depression, and a pervasive anxiety about their future roles in a consumer society.

In the Long Emergency, this scale of educational enterprise will no longer be feasible, and the attenuation of childhood no longer affordable. But the system as it currently exists may be unreformable. For one thing, the psychology of previous investment will weigh heavily on school districts that have built elaborate facilities. They will not surrender to circumstance until it is simply no longer possible to carry on, meaning there is not likely to be any planning or preparation for change. Any effort to reorganize schooling would likely follow a fairly comprehensive collapse of the current system, and would probably occur on a haphazard local basis—perhaps evolving out of home schooling groups. There will certainly be a need for capable teachers, and in a world of vanishing occupational niches, teaching may be considered a desirable job with more social status than it currently offers.

In the Long Emergency, many colleges and universities may close down, and the scale of those that remain may have to contract severely. College will simply cease to be the mass "consumer" activity it has become. The states will not be able to support campuses that have grown to the size of suburban industrial parks; the radically changing job market won't require masses of graduates and postgrads; and the public may be too economically distressed to afford college under any conceivable arrangements. If college in some reduced form remains available to an elite, it might generate political grievance among those foreclosed from the kind of opportunities that earlier generations enjoyed. Depending on the amount of social disorder in the United States, or in particular regions of the nation, those elites might not want to self-identify by congregating in colleges for a while, in which case we may be in for a period when college ceases to exist altogether. Or higher education may become the province of religious sects, as it has before in history.

Higher education is directly related, of course, to the nature of work. In the Long Emergency, vocational trades requiring real skills (farming, animal husbandry, fine carpentry) may gain in status and other professions (lawyering, real estate sales) may lose status and earning power. Some occupations (public relations, marketing, travel agentry, authoring books) may shrink or disappear altogether. Work for many may become a matter of making oneself useful to others in the immediate community, with the possible added benefit of earning a living by doing so. Above all, however, the pervasive and corrosive idea of just being another wage-earning "unit" in a consumer society will be dead. Social and economic roles, long detached from each other within the context of a real community, will necessarily become reintegrated as most work becomes local and more or less visible. Social responsibility to the community will be hard to evade, and status will be based on values other than just annual income.

As these new social arrangements sort themselves out, the Long Emergency is going to produce large numbers of economic losers. The potential for their losses and grievances to translate into political hardship is the subject of the next section.

The Regional Outlook: Sunset in the Sunbelt

It would be reasonable to wonder whether the United States will continue to exist as a unified entity, and what kind of strife the Long Emergency could ignite region by region. Some parts of the nation will do markedly better and others worse—in proportion, I think, to how much they have benefited from the easy conditions of the past fifty years. The American West, especially the Southwest, may suffer inordinately for several reasons. Southern California, Arizona, New Mexico, Nevada, parts of Texas, Utah, and Colorado have been made habitable solely because of cheap energy. Practically all settlement in this region has occurred during the 150-year run of the oil age, with the most explosive growth phase only in the past fifty years. Phoenix and Las Vegas were tank towns before World War II. During this brief and anomalous period of history the region has exceeded its natural carrying capacity to such an extreme degree that even mild to moderate disruptions in the energy supply system will be disastrous. Transportation, air conditioning, and water distribution will become critically problematic in the years ahead. As oil- and gas-based agriculture fails, and it becomes necessary to grow more food locally, places like Phoenix, Las Vegas, Albuquerque, and Los Angeles will painfully rediscover that they exist in deserts.

The Southwest also faces increasing friction with adjoining Mexico. This is not a racist provocation but a description of reality. No other first-world country has such an extensive land frontier with a third-world country. The income gap between the United States and Mexico is greater than that between any other two contiguous countries in the world. "Contiguity," Samuel Huntington writes, "enables Mexican immigrants to remain in intimate contact with their families, friends, and home localities in Mexico as no other immigrants have been able to do."[4] This has provided incentives to avoid assimilating into mainstream American culture and, in fact, has set up a kind of cultural competition that inevitably becomes a contest for political hegemony, especially as living standards fall in the United States.

4. Samuel Huntington, "The Hispanic Challenge," *Foreign Policy*, March-April 2004.

If the American economy loses traction, Mexico will suffer by another order of magnitude, because the health of its economy is so closely linked to ours. As this occurs, the chance for political turmoil in Mexico will increase. During the Mexican Revolution and its attendant civil struggles that persisted from 1911 to 1940, nearly 10 percent of Mexico's population eloped into the United States. The population of Mexico was under 15 million in 1920. In 2005 Mexico's population will reach 106 million. According to the 2000 census, there were more than 4.8 million illegal Mexican immigrants in the United States out of a total of 8 million legal and illegal immigrants combined. Mexicans comprised almost a third of all 28 million Hispanic immigrants in the United States, and the proportion was increasing steadily. The Mexican immigrant population is highly concentrated, with 78 percent living in just four states, and nearly half living in California alone.

The affluence created in the final decades of the cheap-oil blowout made the United States, and southern California in particular, an irresistible objective for Mexican immigration. Jobs were plentiful and wages, compared with those in Mexico, were high. The U.S.-Mexico border today is under only partial control at best. At what point does illegal immigration become extremely undesirable, perhaps even intolerable? If there is such a point—and some Americans would deny that there is—would the United States have to defend its southern border? If so, how will the U.S. government defend this border in a time when the American military is apt to be overcommitted in the Middle East and elsewhere? Ultra-right-wing militias have already sprung up along the U.S. side of the border, manned by exactly the new class of economic losers who will be increasing in numbers as the Long Emergency deepens—poorly educated, underemployed, angry whites. A state of violent anarchy may grip America's southwestern states as militias array themselves against illegal immigrants. And at some point, the immigrants could certainly arm themselves in kind. There is no shortage of small arms in the world, even among the most impoverished societies, as recent experience has shown. At what point does this border conflict turn into a de facto border war?

The *reconquista* or Aztlan movement among Mexican nationals living in the United States cannot be dismissed as racist political paranoia, either. (Aztlan: the legendary ancestral homeland of the Aztecs, which they left in journeying southward to found their capital, Tenochtitlan.) *Reconquista* is established in the streets and in the universities. Charles Truxillo, a professor of Chicano studies at the University of New Mexico, declared in 2000 that a "Republic of the North" should be brought into existence "by any means necessary." He proposed that the "inevitable" creation of a sovereign Chicano nation would comprise the present states of California, Arizona, New Mexico, Texas, and part of Colorado. The belief that this territory was stolen in the nineteenth century is supported unofficially by the Mexican government, which effectively does nothing to stem the flood of immigrants going north—and actually benefits from it as it acts as a safety valve on Mexico's own internal political problems over pervasive poverty.

The acronym MEChA stands for *Movimiento Estudiantil Chicano de Aztlan* or "Chicano Student Movement of Aztlan." MEChA is an explicitly separatist organization that encourages anti-American activities and civil disobedience on behalf of "*la raza*" (i.e., the brown race). Other related separatist groups go under the names of Brown Berets de Aztlan, OLA (Organization for the Liberation of Aztlan), La Raza Unida Party, and the Nation of Aztlan. Some of them mix elements of criminal gangsterism with politics. Although the activism of these organizations varies from somewhat radical to extremely radical, they share the same objectives, the "liberation of Aztlan," as the constitution of MEChA puts it. These groups are firmly established in the barrios of Los Angeles and other southwestern cities. The MEChA symbol is an eagle clutching a machete and a stick of dynamite with a burning fuse.

The reaction so far has mostly come in the form of ballot measures to clarify and enforce entitlement distinctions between citizens and illegal immigrants. Californians have passed initiatives against the granting of driver's licenses and social services benefits to illegal immigrants and to terminate bilingual education. In 2004, voters in Arizona passed Proposition 200, requiring new voters to offer proof of citizenship and state and local governments to verify legal residency before extending "public benefits" to people who apply.

As massive Mexican immigration into the southwestern United States continues to occur in a climate of increasing turbulence and desperation, it will prove to be a tragic and quixotic historical event because the inevitable falloff in oil and gas supplies will drastically reduce the carrying capacity of the "Aztlan" region for all human life, whatever its race or national origin. Two groups will be fighting for control over territory that will be unable to support either. It could take decades for that tragic scenario to play out. In the meantime, we can expect mounting friction between the newcomers and autonomous American paramilitaries, with the federal government rendered ineffective to deal with either immigration or the violent reaction to it.

Trouble is likely to occur in stages. Even as the Long Emergency begins to exert its exigencies, immigration from Mexico will continue and probably accelerate as economic problems afflicting the United States are magnified in Mexico. Mexicans will follow the established pattern of looking to *El Norte* in hard times. This torrent of newcomers will find themselves in a rapidly sinking U.S. economy. Many of the jobs they had hoped to find will have evaporated. With the suburban housing industry crippled, there will be no need for more sheet-rockers. The better-off Anglo population will be losing their jobs, too, and they will not have the cash to employ so many Mexican maids and gardeners.

Those who had lawns may be unable to continue watering them. The water supply situation in the Southwest is likely to become only more critical. Global climate change is already producing extraordinary conditions in the region. In June 2004, officials were calling a persistent western drought the worst in five hundred years. Scientists from the U.S. Geological Survey said the effect on the Colorado River basin was worse than it had been in the Dust Bowl of the 1930s.[5]

In the early stages of the Long Emergency in the Southwest, an increased influx of Mexican immigrants will confront a desperate established population of Americans who, for one reason or another, will be stuck where they are. Many of them will be reluctant to lose the assumed value invested in their suburban houses, even if that mode of existence is

5. The report said the drought has produced the lowest flow in the Colorado River on record, with an adjusted annual average flow of only 5.4 million acre-feet at Lees Ferry, Arizona, during the period 2001–2003. By comparison, during the Dust Bowl years, between 1930 and 1937, the annual flow averaged about 10.2 million acre-feet. From CNN-Online, June 18, 2004.

becoming increasingly untenable. In any case, it is human nature to consider a place "home" if you were born there, or have family there, or have spent some portion of your life there, and people are naturally reluctant to leave home. I daresay that many Americans now living in the Southwest will not be disposed to understand what is really happening—that the carrying capacity of their home region has been suddenly and drastically reduced—and they will hunker down hoping for a return to better times. Those better times will not come back because the conditions that made them possible will be over. A region built on the conquest of vast distances by the automobile, the conquest of unbearable heat by air conditioning, and the conquest of thirst by heroic water diversion projects will find itself hot, thirsty, and stranded.

The vested owners of all those sun-drenched tract houses may stick around for a while and fight over the region, perhaps thinking that they are reenacting the great historical dramas of the nineteenth century—such is the long-term effect of canned entertainment on the collective imagination. The violence and loss of resources will surely send some American citizens fleeing. After a while, it will be obvious to even the staunch defenders that places like Los Angeles, Las Vegas, Phoenix, Tucson, and Albuquerque will never again support the populations that were possible during the height of the cheap-oil blowoff in the late twentieth century. They will then pack up and move elsewhere, sacrificing their houses and ties to a disintegrating community. These new refugees may move into careers that they never could have conceived of twenty years earlier, when they were young college graduates: farmer, farm laborer. Wherever they go, they are going to discover a nation preoccupied with food production above all other activities.

Deeper into the twenty-first century, the region once known as Aztlan and more recently known as the southeastern United States will revert to being a barely habitable arid scrubland filled with abandoned tract housing, deserted freeways, vacated strip malls, decommissioned fast-food emporiums, and all the rest of the equipment that could be of use only in a cheap-energy economy. Whether the region is nominally under the jurisdiction of the Mexican government or the United States may not matter very much. All but the last stragglers will have left, just as the Aztecs did around A.D. 1100. In Las Vegas, the "excitement" will be over.

The Land of NASCAR

The Southwest is sometimes referred to as the "Dry Sunbelt." The "Wet Sunbelt" is the Southeast, the old Confederacy, the states between the Atlantic Ocean and the arid reaches of West Texas. This region has its own set of problems that will make for a difficult adjustment in the Long Emergency. Before World War II, the states of the Old South made up an agricultural region with few cities of consequence. Georgia, Alabama, and the rest were geographically vast states. You could fit twelve Connecticuts inside Georgia and still have room left for a Massachusetts or two. The towns were well served by railroads, but the stretches between the towns were long and lonely, and before the automobile came along, the oppression of distance in the South was as palpable as the summer heat. Except for a few cosmopolitan outposts on the seacoast—New Orleans, Charleston—the region long remained intractably provincial, sunk in virtual serfdom, despotic politics, and superstitious religion.

The technological marvels of the twentieth century began to change that. The automobile especially was a boon. It conquered the demoralizing rural distances between the railroad whistlestops. It is not an accident that the South is a region in which automobile worship has been elevated nearly to the status of a religion, as embodied in the mass communal raptures over NASCAR—professional stock car racing. The popular myth, which Tom Wolfe helped spread in his 1965 essay "The Last American Hero," holds that stock car racing grew out of the exploits of Appalachian bootleggers who souped up their cars to outrun federal revenue agents. Perhaps, but the bootlegger myth tends to obscure the more mundane reality of southern life: that a lot of ordinary people were trapped where they lived by economics and geography and that the transformative influence of the car affected everybody in the South. Driving became synonymous with liberation in the broadest sense. The car was to poor southern whites what emancipation had been for the slaves—jubilee. The car, and a few other innovations, turned the Wet Sunbelt from a mostly agricultural backwater into a late-twentieth-century corporate industrial economic dynamo.

Those other innovations were, of course, rural electrification, which lagged behind the introduction of the automobile by a good two decades, and universal air conditioning, which arrived much later. Electrification in the northern states had proceeded earlier and much more expeditiously between 1890 and 1920 because the northern states were proportionately much more urban. Customers, including big industrial users, were concentrated in cities and towns. Electricity could be generated in urban power plants (using coal) and distributed through dense local grids. Rural electrification in the northern states occurred with the installation of interurban trolley lines, which were designed largely for moving workers to factories. The factories themselves, when located on waterways, were often capable of generating their own electricity. Northern towns also built hydroelectric power stations at what today would be considered a very small scale of operation.

Electrification in the South happened differently. In 1900, the South had few cities, not many factories, and many agricultural trading towns without the means to supply their own power. The enormous flat expanses afforded few exploitable hydroelectric sites. Lack of electricity across vast reaches of the South kept the region mired in backwardness through the 1920s—just when a depression in farm commodity prices occurred as a result of mechanization. Electrification of the South was therefore a mostly rural project; it was accomplished by harnessing water power in the rugged Tennessee and Cumberland river systems, and sending that power to sparsely settled customers over great distances. The required infrastructure alone in the form of dams, transmission lines, towers, and relay stations was so formidable that it required large federal government subsidies to make it happen. The electrification of the rural South was not begun until the 1930s, with the federally sponsored projects of the Tennessee Valley Authority (TVA). The scale was fantastic. No sooner were they completed than World War II had commenced. When it was over, the South was ready to launch itself as a new economic phenomenon. The elements were all in place: cheap distributed electricity and cheap oil, meaning that geography had lost its constraints. Air conditioning went first into public places—theaters, restaurants, and department stores—in the 1950s, and finally into middle-class homes on a widespread basis in the 1960s.

The result was a region that would base its economy largely on suburban expansion. Cheap land was plentiful across the South and the region was culturally predisposed to an antiurban bias. In the minds of most developers, suburban development was the logical, natural way of improving on country life. The fact that it decimated rural landscapes and rural lifeways didn't seem to matter, because by the mid-twentieth century in the American southland, rural life was more about being in cars (or trucks) than growing cotton, with all the horrors of serfdom implied by that form of agriculture. If anything, postwar southerners tried to forget what rural life had really been about for them and their ancestors, and manufactured a sentimental version to replace it, which was later sold back to them in the form of commodities such as popular music, corporate religion, and theme park admissions.

Nashville is a good example of how this mentality worked. Prior to World War II, Nashville had some legitimate claim to being if not exactly a great city, then a large town of consequence. It called itself "the Athens of the South." It had a full-sized replica of the Parthenon in one of its parks and a world-class university (Vanderbilt) with a cultivated international faculty. As the region became more affluent in the 1960s, country music became an important "industry," with Nashville designated its "capital." In order to thrive economically, Nashville had to identify itself with everything "country." This meant, paradoxically, repudiating its qualities as a town. Nashville therefore did an excellent job of destroying most of its center in an orgy of late-twentieth-century "urban renewal," while the stars of country music settled in suburban villas outside of town, that is, in the "country," along with their fans, who didn't want to have anything to do with the town, whose remaining inhabitants were predominately the descendants of deracinated African American sharecroppers.

Socially, suburbanization represented a decisive victory over everything thought to be represented by city life, especially things thought to be effete and unmanly, like the traditional arts. Of course the ersatz country folk of suburbia had their own art, country music, but wearing large hats immunized its practitioners against effeteness and made it manly. Eventually, its holiest shrine, the Grand Ole Opry, moved out of its downtown auditorium into a cheapjack plastic theme park in the suburban hinterlands.

Southern small-town and rural life had been largely a culture of poverty for whites from post-Civil War reconstruction until the 1950s, when the broad southern cracker lumpenproletariat finally began to join the middle class for the first time. Having come relatively late to this region, the wonders of twentieth-century technology were embraced with unalloyed exuberance. After air conditioning became widely affordable, southerners hardly went outside anymore, unless it was in a motor vehicle. Anything about southern vernacular architecture that once had been graceful in adapting to the climate was cast aside for the pleasures of air conditioning and cheapness of construction. The new southern middle class suddenly had a lot of money to spend, but they spent it all on plastic junk and anything with a motor in it. They got most of their ideas about the world from movies and television. Suburbia was their dream habitat. It was the "country" with all the dust, heat, toil, and flies removed.

The suburbanization of the South was certainly augmented by the relocation of industries from the wintry North to the balmier South. In places such as Charlotte and Memphis people could throw a baseball around outside on the lawn in January, and the daffodils poked out of the ground a month later. The summers were frequently unbearable, but the new nature of work—indoor corporate office work—took place in air-conditioned office park buildings, so people barely noticed the bad months. Anyway the economy of the so-called New South wasn't so much about the corporations themselves relocating to places like Atlanta and Orlando. They only supplied a customer base for the real economy of the New South, which was real estate development, the creation of new reaches of suburbia, with the housing pods, miles of limited-access highway (at an average cost of $20 million a mile), and the construction of accessory highway strips, regional malls, big-box stores, "power centers," office parks, and everything else required for a drive-in utopia.

An additional factor was the influx of late-twentieth-century retirees to states where snow and ice were seldom seen. The retirees of the late twentieth century were the most coddled generation of old people in world history. They benefited tremendously, in the form of pensions and asset appreciation, from the accumulated wealth of the cheap-oil fiesta that had gone on their whole lives. Most of them, too, had lived their whole lives in thrall to the automobile. They had been born in the morning of the auto age and enjoyed its heyday, and now found their mobility magically extended into old age. The elderly were the staunchest supporters of the suburban way of life, and its greatest beneficiaries—just as children were the biggest losers in a culture of extreme car dependency. Seniors voted in greater numbers than any other group, they sat on the planning and zoning boards of most localities, and they brought an extreme pro-automobile bias to every decision entailed in the design of the built environment. By doing so, they conditioned the habits and practices of the house-building and retail industries, so the result was a human habitat designed primarily to be connected by cars—and virtually useless otherwise. The proclivities of the elderly found eager allies in the southern middle class, who desperately wanted to dwell in a world of comfortable climate-controlled, artificial places.

This is what came to be the overwhelming physical reality of life in the postwar South and this is what southerners are now stuck with. In the Long Emergency, not only will a further elaboration of suburbia be impossible—and therefore an economy based on that activity—but what is already there will lose value and usefulness practically overnight. This, of course, will happen all over the United States to one degree or another, but the consequences in the Old South will be worse. The broad themes of cracker culture, still very much alive in the middle class, include a marked tolerance for violence, even a glorification of it, as seen, for instance, in the regional strength of the National Rifle Association, the "outlaw" mythology of country music, and the disproportionate numbers of southerners in the armed forces.[6]

It is true that the middle class in metro-burbs like Atlanta has been culturally diluted by an influx of people from other parts of the country, and even other nations, but a significant residue of the original stock remains, with all its latent encoded behavior in place. As suburbia fails in an oil-challenged world, this middle class will lose its entitlements to a comfortable way of life—to large air-conditioned houses beyond walking distance of anything, to secure employment, to easy motoring, bargain shopping at the Wal-Mart, and cheap, mass-produced food products.

6. "A Spanish official reported the 'influx [into Florida] of rootless people called Crackers.' He described them as 'rude and nomadic, excellent hunters but indifferent farmers who planted only a few patches of corn, as people who kept themselves beyond the reach of all civilized law.'" (OED) "A German visiting the Carolina backcountry found longhorn cattle, swine, and slovenly people whom he identified as 'Crackers.'" Grady McWhiney, *Cracker Culture: Celtic Ways in the Old South*, Tuscaloosa, AL: University of Alabama Press, 1988.

This group of people will be very angry and bewildered at the loss of their entitlements—only so recently acquired—and they will express their anger both politically and extralegally. Politically, the southern middle class is the most solidly conservative of any group in the nation. What "conservative" actually means anymore may be subject to debate, but for the sake of this argument let's say it includes the following attitudes: (1) that the United States is an exceptional nation whose citizens enjoy special dispensations from a Christian god for their good works in being a beacon of liberty for the rest of the world; (2) that being an American is characterized by rugged individualism that tends to exalt the family, clan, or tribe while it discounts the common good of the community and even the rights of others not of the family, clan, or tribe; and (3) that firearms should be liberally distributed among the populace with the expectation that they will be used to defend individual liberty. Anyone can see that against a background of grave economic distress, attitudes like these could lead to a great deal of delusional thinking, dangerous politics, and possibly mayhem.

The Old South, like other regions of the United States, will face the necessity of building a new economy and radically reorganizing its land-use practices in order to create that economy, which will surely have to put food production at its center. Both elements of this enormous task would be daunting under the most favorable circumstances, let alone under conditions of political grievance, paranoia, and turmoil. The project of suburbia was itself a radical experiment in the reassignment of land uses, and much of what had been good farmland until a few decades ago has been built on and paved over, especially land at the edge of towns and cities that had been the proximate source of food supply until the era of continental-scale industrial agriculture. Those suburban housing tracts and big-box stores are not going to be disassembled or moved any time soon, if ever. Can families feed themselves on what they grow on a half-acre lot? The idea of suburbanites in such numbers as found around Atlanta all resorting to subsistence gardening on a successful basis seems unlikely. The majority may not even try. Desperation might provoke movements of unprecedented numbers of people away from the southern suburbs and the metroplexes to what remains of the rural landscape across these large states. There is still a lot of undeveloped land there. But how do you reallocate this land? Could the process remain orderly and peaceable, given the number of small arms available and the regional cultural predisposition to violence?

Land that has been in agricultural production during the past fifty years has almost certainly been farmed in the industrial mode. If so, the soil may be organically "dead" after decades of liberal chemical pesticide use, meaning it will not contain the complex communities of microbes needed to turn organic tillage and manure into humus in a self-sustaining way. The soil also may be depleted of vital minerals, which previously had been compensated for with decades of artificial fertilizers. In any case, the health of the soil will be hard to restore and before it is accomplished the people may have to endure very hard conditions. We also don't know what the effects of global warming and climate change will mean in the American Southeast. It may become more subtropical. But it might also suffer decreased rainfall and prolonged drought. What kind of crops might be grown there? If diseases found in the Caribbean region extend their range into the Southeast, farm labor may take on a high mortality rate. But human life may also be a lot cheaper than it is now.

Even more troubling than the logistical difficulties are the social implications. There are two previous models for farming in the American South. The first was the plantation system based on slavery. The second was really a modified version of the first when slavery was outlawed: sharecropping, in effect, serfdom. Both systems are essentially feudal. I doubt slavery will make a comeback, but I wonder about sharecropping, or something like it. The persistence of culture is a real phenomenon. Over a period of time, a southern agricultural economy may reorganize itself by default along the feudal lines that existed historically, odious as it may seem. This may be even more likely if land tenure issues end up being settled by violence and new social class lines are established by pitting the strong against the weak.

The Long Emergency will cause unprecedented social and economic dislocation, and the outcome may be a world we would barely recognize. The relatively egalitarian society we knew in the late twentieth century may become drastically more hierarchical as large numbers of desperate people place themselves in the service of those who control land, especially following a period of anarchy. Under such harsh conditions, the weaker individuals will sell their allegiance in return for security. You can't feed yourself or your family if you are dead.

Christianity Inflamed

Since central government is likely to be increasingly ineffectual and irrelevant in an energy-starved world, I believe the dominant culture of the American South would tend to promote people taking the law into their own hands. Any transition, therefore, from an economy based on suburban sprawl to an economy based on small-scale farming and small-scale commerce serving it will probably be very difficult and disorderly in the southeastern United States. It also seems to me that whatever government, if any, does finally resolve out of the anarchy faced by this region, it may end up being despotic and theocratic. The South to this day evinces levels of extreme religiosity far above other parts of the United States and is, in fact, the home base and spawning ground of many sects of Christian fundamentalism.[7] Pentecostal or evangelical fundamentalism has long been integral to the politics of the region. As the postwar economy uprooted so many southerners from rural places and traditional ways of life, and plunked them in alienating, lonely, disconnected suburban nowheres ruled by consumerist ways of life, religion became ever more important as the only remaining place of social enactment. Church membership across this sterile suburban social landscape increasingly compensated for the absence of real communities based on networks of local economic relations. In a way, fundamentalist religion made the corporate predations of community-destroyers easier. It made secular community seem optional, dispensable, provisional, something easily replaced by Wal-Mart. It squared nicely with the ethos of hyperindividualism, in which

7. "Between 1960 and 2000, the number of Americans who attended weekly services fell from thirty-eight per cent to twenty-five per cent. At the same time, membership in the Southern Baptist Convention grew from ten million to seventeen million, and membership in the Pentecostal churches from less than two million to nearly twelve million." *The New Yorker*, Books, David Greenberg, "*Fathers and Sons: George W. Bush and His Forebears*, July 12 and 19, 2004, p. 97.

bargain shopping trumped any aspect of civic amenity. The churches, meanwhile, sought to benefit from the same economies of scale as those enjoyed by the giant retail chains. Increasingly, the churches were organized on a mass basis and housed in buildings that looked like Wal-Mart with gigantic parking facilities. In fact, evangelical churches were renowned for taking over the leases of dead chain stores in dying malls because the rents were so cheap. Southern evangelicalism became a kind of Wal-Mart of the spirit. Political leaders went bargain shopping in them for voting souls.

It was also a rather severe belief system, with its emphasis on sin and punishment, which provided a means of social control needed originally among rural poor folk inclined to casual violence. These themes redounded in the stories of judgment and "end times" that often preoccupied fundamentalists, especially those parts of scripture concerned with actual apocalypse, such as the book of Revelations. It is perhaps a strange and unfortunate coincidence that the hardships presented by the Long Emergency will seem to be a playing out of the fundamentalists' most cherished and extreme prophecies. It is not the author's contention that the Long Emergency represents apocalypse, as my argument is based on the continuation of human life and the project of civilization in particular. But to many evangelicals, the end of oil-based comforts and amenities may amount to the end of the world. It will certainly be the end of many habits, practices, assumptions, and ideas. The conditions of the Long Emergency will probably reinforce the belief among the pious that evil forces or persons are behind the troubles of the post-peak oil world, and offer justification for punitive actions taken in the face of those conditions.

Of course, like anything organized now at the giant scale, southern fundamentalist religion may suffer in the Long Emergency. The huge industrial sheds built as churches will be hard to heat—it does get to freezing in Atlanta in January—and cheap air conditioning will cease to exist. Members will not be able to travel very far to attend. Churches will have to scale down by necessity, just as schools will. If they become smaller, there may be more of them, requiring more pastors, and there may be quite a bit of competition, or even friction, among them, with multiplying factions, denominations, and new sects. Organized religion, and especially action taken in its name, may collide also with the individualist ethos that rules at times other than Sunday morning. Some individualists may peg their allegiance not to organized churches of unrelated semistrangers but to family, or extended family in the form of tribes, clans, and gangs struggling for survival in an energy-scarce world. In the early phases of the Long Emergency, individualist forces might compete for power with fundamentalist pseudocommunities. These families, tribes, clans, and gangs may eventually prevail and mutate into a ruling class, coopting religion and enlisting its ministers to pacify a frightened and surly populace, especially as it is needed for farm labor. What I am describing, of course, is an old story, seen historically in many places, from the Catholicism of medieval Europe, to the Christianization of American slaves, to the orthodox church of nineteenth-century Russia. The populace, however, might be fewer in number than today's due to attrition from starvation, epidemic disease, and incidental violence.

The Old Union

The states of New England, the mid-Atlantic, and the upper Midwest, which make up the historic Old Union of the Civil War period, also seem united in destiny as they face the Long Emergency—just as they seem distinct in many ways from the Old South and the Southwest. They will hardly be immune to the ravages of energy resource depletion, but people in the Old Union may be in a somewhat more favorable position to survive and begin to fashion a transformed society capable of carrying on the project of civilization.

For one thing, the region possesses the residue of a preindustrial or protoindustrial civic infrastructure, based on established habits of group covenant and, ultimately, a respect for the rule of law. Physically, the Old Union still contains an underlying fabric of towns and small cities embedded in some of the nation's best agricultural terrain, and the cultural memory for using them as an integral living arrangement still exists. The civilizational software for running democratic communities remains alive. In distinction to the Old South, the latent encoded behavior of the vested authority in the Old Union is derived from the idea of commonweal, as opposed to individual or the clan. Indeed, the idea of vested authority itself remains unsettled in the Old South, except in its feudal atavisms. The institutions for maintaining a commonweal enjoy a more certain and abiding respect in the Old Union, in particular the law and the courts that serve it. Except among mafiosi and gangbangers, there is no generalized habit of settling differences extralegally, and violence is not normative or regarded with romance-tinged tolerance. This is not to say that the region I call the Old Union has no history of bad behavior or endemic collective shortcomings. But I regard it as less likely to fall into hopeless lawlessness, anarchy, or despotism, and more likely to salvage the bits and pieces of our best social traditions and keep them in operation at some level.

The Old Union is far more secular than the Old South, less beset by superstitious and despotic fundamentalist religion, and more generally inclined to observe a traditional separation between church and state. The Puritanism that looms distantly in its history contained a core of discipline that worked in favor of group survival under conditions of extreme hardship that may return in the Long Emergency. Eventually, that Puritanism evolved into the resourceful and disciplined traits we associate with New England Yankees at their best: thrift, rectitude, perseverance, and allegiance to a community. Its cultural baggage, via a line of figures as diverse as Ralph Waldo Emerson, Abraham Lincoln, Theodore Roosevelt, H. L. Mencken, and Camille Paglia, includes a lively conversance with the history of ideas free of dogma and cant.[8]

The Old Union also contains many of the nation's largest industrial cities, as well as the most extensive contiguous suburbs, all presenting substantial liabilities, as outlined earlier. In the short term, the biggest cities will be places of desperation, disorder, and economic loss. New York City could become largely uninhabitable if the electric grid and the natural gas distribution system malfunction even moderately. Boston, Philadelphia, Baltimore, and Washington present similar logistical nightmares, though they contain far fewer high-rise buildings than New York City. All other things being equal, I believe every one of these cities will shed large fractions of their populations as the Long Emergency continues, year after year.

8. I hesitate to add Tom Wolfe, a Virginian (and a classic cavalier personality) who has lived in New York City for forty years.

Even as that happens, they will retain the value of their important sites. Some people will remain in them. New arrangements and economic behaviors will emerge in them, and some healthy elements may begin to take shape, even as the cities continue to experience a generalized agony of contraction. For instance, New York's waterfront may come back into use at a much smaller scale than the city experienced in the previous twentieth-century heyday of gigantic freighters and behemoth passenger liners. We may see the return of sailing ships, or sail in combination with the "twilight" technologies of oil and even coal-based steam. Small-scale commercial fishing may return in the once-bountiful Hudson estuary. A new New York might coalesce around these things on a reactivated waterfront, in a new low-rise community of buildings and small-scale wharves—while in Manhattan's interior, the old corporate towers remain abandoned and dangerous. Similar events might occur in Boston, Philadelphia, and Baltimore.

A process that might transform them into livable, fully functioning, appropriately scaled human habitats might take decades or centuries. The architectural theorist Andres Duany has remarked that the ideal civic environment might be found in the intimate form of the Gothic city accessorized with halogen lights and decent plumbing. The great, brutal, hypertrophic industrial metropoli of America might evolve into much more intimate and human-scaled places. We might consider ourselves lucky if the process gets under way in the twenty-first century.

The gigantic smear of suburbia that runs almost without interruption from north of Boston through Connecticut, New York, New Jersey, Baltimore, Washington, and northern Virginia is not going to be a happy place. It will be subject to the most extreme loss of utility and equity value. This suburban portion of what was once called *megalopolis* may become as much a forbidden zone as the South Bronx became in the late twentieth century. The inhabitants of the Bronx in 1925 would never have believed how desperate their borough would become in 1970; by the same token, the denizens of Bergen County, New Jersey, or Fairfield County, Connecticut, today may never believe how desperate their localities may become in 2025. Of course, the demise of these suburbs will also be the beginning of a transformation to something else. Decades or centuries forward, even Hackensack may return to being a plausible environment—its brief history as an automobile-choked fistula of bargain shopping and free parking all but forgotten. There is extensive territory in the Old Union beyond this vast northeastern suburban wasteland that could evolve into some of the relatively more successful localities in the dazed and crippled America of the Long Emergency.

Water resources and water power will be extremely important in the twenty-first century and the Old Union has plenty of water, as well as the topography that makes water follow the law of gravity and become available for useful work. New England has the capacity to reinstate a comprehensive system of small, community-oriented hydropower operations, as well as small-scale manufacturing based on hydro. Its agricultural potential is greater than it currently appears, based on today's sclerotic diary industry. The region could go a long way toward feeding itself if it had to. Western New York, alone, could be far more productive. Amish families have only recently begun to expand there from Pennsylvania, and their example to others in the use of traditional methods can only be salutary as the Long Emergency advances. Northern New York along the broad St. Lawrence frontier was once the cheese capital of the Northeast. Ohio, Indiana, Michigan, Illinois, Wisconsin, Minnesota, and Iowa are all capable of renewing local economies based on productive small-scale farming serviced by small towns and regulated by habits of civic behavior and the rule of law. These states will have fewer resources for hydropower generation, however. And in the absence of nuclear power they may be comparatively shorter of amenities.

The Great Lakes have been a deeply underused economic resource for the past half century. This remarkable freshwater inland sea stretching from New York to Minnesota has the potential of unifying an ordered matrix of towns and farms and appropriately scaled cities with a transportation system that is not dependent on nonrenewable energy. Without the destructive, entropic effect of cheap oil, a Great Lakes fishery might become both productive and sustainable. The small towns are still there, waiting to be rehabilitated—mothballed at their centers and encrusted with strip malls on their edges. They will be less difficult to reconfigure than either the great cities or the suburbs. Chicago and its sprawling outlands will present the same liabilities as New York. Detroit and Cleveland, however, are both so far gone in contraction and abandonment that they might skip some stages in the process of recreating themselves. Both enjoy strategic sites on bodies of water. They will surely be smaller than they once were. Conditions are similarly far gone in Dayton, Akron, Toledo, Columbus, Flint, Grand Rapids, Indianapolis, Des Moines, and Davenport. They are already crippled and abandoned and none are so big that size alone would prevent them from reconstituting themselves. Minneapolis and St. Paul, like Chicago, will have to endure a more substantial meltdown of their highly suburbanized precincts. Pittsburgh enjoys a favorable site at the Ohio River junction of two other rivers and could become again what it once was: a water transport gateway into the Midwest.

In recent years there has been talk of a colossal pipeline project that would send monumental quantities of fresh water from the Great Lakes to rescue the Southwest—thus allowing the continued suburban expansion of cities such as Denver, Phoenix, and Las Vegas. Since it is the premise of this book that continued suburban expansion is impossible, I don't think this has to be a concern, not to mention the fact that a project of such scale would also violate the principal terms of the Long Emergency, which will obviate all super-scale enterprise. It is conceivable that the Great Lakes states of the Old Union might merge politically with the Canadian province of Ontario at some point in the future, as they may have more administrative interests in common with each other than with the other regions of the United States or Canada.

The West

By the West I mean the Great Plains and the Rocky Mountain states. The outlook for them in the Long Emergency is dismal—or, more specifically, they are destined to become among the most depopulated, unproductive, and desolate places in the nation. This process was well under way in the Dakotas, Montana, Nebraska, western Kansas, Colorado, and Oklahoma by the 1970s. Farming the semiarid Great Plains may prove to have been an experiment that failed. It has gone on for only a little more than a century. Farming was established in this inhospitable region largely because the land was given away free by the government to anyone who would try to farm it, and many of those who tried were immigrants who came there with nothing to lose. The experiment was already failing in the late nineteenth century when oil and mechanization came along to keep it going by artificial means—they allowed the cultivation and irrigation of much larger acreages by far fewer people. Government subsidies and corporate exploitation of industrial production methods have also kept Great Plains farming on life support for decades. Some of the western counties of Nebraska today are the poorest in the nation per capita, and there are fewer and fewer capita each year. Whole towns out on the prairie are drying up and blowing away. Even the Wal-Marts won't go there. There has even been talk lately of merging North and South Dakota to save on the administrative costs. The Long Emergency will put an end to farming in most of these places. The region will revert to being a sparsely populated backwater. Before the century is over, most of the dams in the extensive Missouri River system will have silted up and seasonal flooding will resume, making life more difficult in the remaining towns. Proposals were made by environmentalists in the late twentieth century to programmatically return broad expanses of the Great Plains to their natural condition, short grass prairie, complete with roaming buffalo and antelope playing. The creation of this "buffalo common" may not now require any conscious effort or program. It will probably happen naturally as industrial farming ceases and more people leave. With automobile use on the wane, the region will become less accessible. Its extremely monotonous landscape and horrible weather will not lure many visitors. Indians, and groups behaving in the mode of Indians, may eventually resume traditional ways of life there.

A similar, perhaps even more desolate fate, awaits the Rocky Mountain region. Idaho, Montana, and Colorado have become magnets for a new type of yuppie hypersuburbanite in recent decades, many of them sports and outdoor adventure enthusiasts from California and the East who have been able to compensate for the forbidding conditions of their mountain homes with cheap gas, four-wheel-drive mega-cars, long commutes, snazzy technical gear, and airplane tickets to better climes when the weather gets them down. This high-entropy way of life masquerading as sporty environmentalism will be impossible in the Long Emergency. The outdoor adventures will not have to be contrived. Anyone choosing to live thirty miles from a store had better adopt the lifestyle of either a homesteader or a trapper, and both careers are likely to have a high attrition rate. The Rocky Mountains present two other practical problems for permanent residents: Very little food can be grown there, and much of the mountain region is arid, despite the snow-capped peaks.

There are only a couple of cities of consequence in this region. Denver may resume its role as the regional railhead for livestock, but mining, its other original reason for existence, might be played out. Meanwhile, its enormous suburbs will be subject to catastrophic contraction and the whole metroplex could sink into lawlessness. Denver's proximity to the potentially contested territory of the Southwest is also problematic. Salt Lake City, and indeed the habitable valleys of Utah, are peopled by a religious sect well scripted for vicissitude. Families of the Church of Jesus Christ of Latter-Day Saints, or Mormons, are required by their theology to keep a year's food supply in storage in the event of a disaster. Mormon churches are still rigorously ordered into local wards, and members' lives are highly regimented by a hierarchical command system. These habits of behavior might dispose them to survive the hardships of the Long Emergency. But they also have the highest birthrate of any group in America and have far exceeded the natural carrying capacity of their adopted homeland. Salt Lake City is among the most car-dependent places in America, Its current incarnation as a dynamo of western business is completely underwritten by cheap oil and gas, to which Mormons have become addicted like the rest of "gentile" America. When the supply becomes irregular, Salt Lake City will shrivel and some Mormons may find themselves on the march again.

The Rocky Mountain region probably contains more political extremists and cryptoreligious zealots per capita than any other region of the country. Its remote valleys and mountain keeps are home to a breed of super-individualists who abhor authority and harbor paranoid fantasies about Jews, blacks, Catholics, foreigners, and the "New World Order." The Long Emergency will stoke their paranoia and make the places that they control extremely dangerous. Though many fancy themselves survivalists, they will discover the hard way how dependent they actually are on fossil fuels and high technology, and within a relatively short period of time, an inability to grow food will drive all but a few out of the mountains.

The Pacific Northwest

My definition of this region is limited to coastal Oregon and Washington, northern California and lower coastal British Columbia (though currently it is part of Canada). Other commentators reflecting on America's post-cheap-oil future often view this region optimistically. The climate is certainly very favorable and it contains some of the very finest agricultural land in North America. Its landscape is beautiful and the coastal area is well-watered. Both Seattle and Vancouver are gateways to the ocean and might conceivably support downscaled sustainable fishing economies. In their current condition, however, they are typical urban hypertrophies with gigantic suburban appendages, and they will be subject to the same agonies of contraction as other less-picturesque North American cities. It remains to be seen how much of northern California might be infected by the socioeconomic meltdown of southern California. The Bay area will surely be troubled by the extent of its suburban sprawl, and the Central Valley as far north as Redding may be overwhelmed by refugees from Mexico and the additional problems they will create. Oregon and Washington are large states, but much of their terrain is high desert and will not support sizable settlements. Both states may be overwhelmed additionally by sheer numbers of Californians fleeing the disorders there.

The Northwest may find itself in a whole other strange kind of trouble. Exposed to the Pacific, the region may be molested by military or paramilitary seaborne adventurers originating from the far side of the Pacific rim. In the Long Emergency, Asia will find itself in turmoil at least as severe as anything happening in North America, and quite possibly a lot worse because of its swollen populations. China, Japan, Korea, Indonesia, and Malaysia are all populated by seafaring people. As resource scarcity intensifies, and international conflict with it, and America loses its ability to dominate every corner of the globe, Asian navies might reach as far as North America seeking the means of survival. If the Asian nation-states fail, or dissolve into anarchy, restive populations may circulate far beyond the crowded shores of these places in ships and boats. A new kind of piracy could make life in the Pacific Northwest of North America difficult for a long time to come.

To sum up the regional picture, every part of the United States will be tested by the Long Emergency, but some places will do better than others. The Southwest will find itself in the most trouble, as its existence is predicated solely on cheap energy and imported water and can produce no food of its own without them. It will also suffer from proximity to a destabilized Mexico. The Old South faces the revocation of high living standards obtained relatively recently and will have to contend with religious despotism and a latent culture of violence while reorganizing a hypersuburban economy into what may amount to agricultural neofeudalism. The Old Union begins with advantages in water supply, good farmland, and a predisposition to civic habits, but will be burdened with the albatrosses of cratering megacities and megasuburbs. The High Plains and Mountain West region is destined to hemorrhage population and will afford few opportunities for resettlement under post-oil terms. The Pacific Northwest's benefits of mild climate, abundant water, and good farmland may be overwhelmed by populations fleeing the problems of southern California, and compromised by exposure to desultory Asian aggression along the seacoast. The federal government may lose its ability to govern the nation as a whole effectively and the regions may resort by default to autarky. In any case, life in all these places will be intensely local and success or failure will depend on the quality of each community.

Racial Conflict in the Long Emergency

The question of race relations naturally presents itself in any meditation on America's cities and their future. By this I mean relations between whites and African Americans specifically—and from here I will refer to African Americans as blacks for the sake of brevity. While it might make some readers uncomfortable, it would be irresponsible for the author to duck the issue. The public discussion about race relations has been disingenuous during the still-ongoing era of political correctness. Political correctness itself came about largely as a defense against the partial failure of the social justice project of the late twentieth century. It is probably useful to begin by describing what has happened recently before turning to where things stand now.

The civil rights movement of the 1960s had been primarily about removing the legal obstacles to full participation for blacks in mainstream American society dominated by whites. The presumption by educated "progressives" through the 1970s was that once legal barriers came down, blacks would seek to participate and assimilate into mainstream culture. This assumption must have seemed reasonable at the time, given the obvious earnestness of the era's black leadership—Martin Luther King Jr., A. Philip Randolph, Thurgood Marshall, James Farmer, et al.—but the assumption proved to be wrong as times rapidly changed.

When the legal barriers came down, enough blacks demurred in assimilating culturally to create a crisis in the civil rights movement. There were two responses. One was the creation of extralegal cultural remedies of the type known as affirmative action in order to further stimulate assimilation. The other response was black separatism, a simple opting out of the need to assimilate. Malcolm X was the avatar of a new separatism, and a martyr to it. The "Black Power" movement gained traction after his 1965 assassination, paradoxically, just after the passage of the most sweeping federal legislation mandating equality before the law since the Civil War. That Black Power seemed romantic, sexy, and glamorous tended to conceal its retrograde impact.

As a practical matter, Black Power led quickly to some very counterproductive collective behavior. One was the officially sanctioned resegregation of facilities only recently desegregated. By 1970—five years after Congress put an end to "Jim Crow" laws—college administrators were caving in to demands by militant student groups to charter separate black student unions and separate dormitories. Black studies followed as a separate academic discipline. This set in motion a debilitating ethos that persists to this day—the complaint that a "structurally racist" white-dominated society prevents blacks from full participation in a culture many have already opted out of. White "progressives" have tragically supported this ideology, along with a repudiation of mainstream culture itself in order to discount the value of assimilating into it in the first place. The result has been extremely unfortunate.

While it is true that many blacks have joined the middle class, at least in terms of jobs and pay, a disturbing aura of cultural separatism persists, supported by the multiculturalists in education, with terribly demoralizing effects on that substantial minority of the minority who never made it into the middle class. For the past two decades, lower-class blacks especially have been encouraged only to become more separate, more different in behavior, more divorced from mainstream norms of speech, manners, and costume. This dislocation is reflected ominously in pop music. Hip-hop has to be taken seriously because it is so pervasive, and it presents a range of compelling cultural meanings. The most threatening, of course, is its association with criminal behavior—the rhetoric of gangsterism, the glorification of gunplay and murder, and the grandiose imagery of unearned riches. Street mythology has it that hip-hop clothes, accessories, and lingo are extensions of jailhouse fashion. Less obvious is how much these childish conventions of manner—exaggerated clumsy body language, pants many times too large, hats worn sideways—infantilize their followers. Children do not engage in politics, and so one of the worst aspects of this sector of pop culture has been the wholesale depoliticization of the black population, especially young adults. Another result of this surrender of politics to entertainment has been an amazing dearth of black political leadership at a time when it couldn't be more desperately needed to resolve the unfinished business of the social justice project.

There are real political issues facing the black underclass minority in America, and the outstanding one would seem to be how much longer significant numbers of them can afford to put off growing up. The twenty-year-long peak oil blowoff has made this experiment in arrested development possible. If nothing else, it has kept enough surplus wealth sloshing through the economy to keep the party going. The Long Emergency will force the issue. No group of Americans will be able to party through it. Even among the nominally poor today, standards of living have a long way to fall. What remains of the post-welfare reform social safety net may unravel altogether.

The grievance and belligerence that smolders under the surface of the hip-hop saturnalia is unattached to any coherent political claims beyond the debatable clichés of "structural racism." But that belligerence is more a fashion statement than a political message. Glowering behind sunglasses in a rap video is a show business convention now, but the stringencies of the Long Emergency will change the way such posturing is interpreted. The Long Emergency will be such a hardship for everybody, of all races and sexes, that claims of prior special grievance will be dismissed. The Long Emergency will demand so much of individuals in terms of personal responsibility, civic cooperation, and adult skills, that large numbers of people will be unprepared to cope, and the rest won't be disposed to excuse the truculence or misbehavior of those who cannot. They will be too busy working to feed themselves and to stay warm. The remaining question is whether this lowered threshold of tolerance will operate within a context of law, or whether the social fabric will be so tattered by hardship and destitution that the mechanisms of justice will no longer be in force.

Since I believe that life in the Long Emergency will become profoundly local, then the answer really depends on how successful a given locality may be at maintaining civic institutions, including the police and the courts, and, of course, how fairly these things might operate. There is liable to be wide variation. We know from historical experience that racial justice has not been well served in the Old South. We might flatter ourselves to think that it has been better served in other parts of the country. It is obvious that the regional demographics have changed. For the past fifty years, lower-class black culture has been identified with inner cities, the result of a "great migration" that sent several waves of southern agricultural serfs north to cities, just as the national economy was well on its way to shedding its manufacturing sector and the good jobs that went with it. The outcome of that has been extremely discouraging for all concerned, both blacks and whites, those trapped in the violent purposelessness of postindustrial city life, and those dispersed to the alienating precincts of the automobile suburbs. But the cities will not remain in their present shape and condition. Since the 1990s, a reverse migration has been under way with northern urban blacks returning to the southeastern states. In many cases they are returning as members of the middle class. Both California and New York saw the largest numbers of these outmigrations. Those left behind in the urban ghettos may find themselves even more economically and culturally isolated as the Long Emergency begins.

At their worst, the rap videos played on cable TV resemble the war chants of a conflict that has not yet been joined. Only among a group as narcissistically lost and clueless as white suburban America would these messages be welcomed as just another species of entertainment. In the disorders of the Long Emergency, when the poor become really poor by world standards, the urban ghettos may explode again, and the next time it happens it will be in the context of a much more desperate society than the one that witnessed the 1992 Rodney King incident and its aftermath. It is unlikely to be confined to the ghettos themselves but will likely resolve into a more generalized and protracted guerilla warfare of the kind that has been going on in third-world countries for decades, and it will occur against a background of widespread turbulence everywhere.

American exceptionalism offers no protection from these potential disorders. Any place can become a Beirut under certain unfavorable circumstances. We can only hope to hear the appeal of those "better angels of our nature" invoked by Lincoln the last time the United States went through an internal convulsion.

Ideas, Morals, and Manners in the Long Emergency

We tend to take for granted that our contemporary ideas are the most highly evolved ever known, and that our attitudes are the most acute and progressive. I think the truth is that our ideas are suited to a particular set of circumstances. Our political ideas, including extreme examples such as Marxism-Leninism and fascism, derive from the material advances of the industrial age and the competition they have spawned in ever-growing populations. Similarly, our cultural ideas are derived from both the benefits and the disastrous side effects of the industrial adventure, for instance modernism in the arts. The idea of human progress that attends these ideas is inherent to the industrial experience, which was itself an outgrowth of the Enlightenment. Confusing these issues is the temptation to believe that the material progress of the last two centuries has provoked some parallel progressive evolution of human nature, making us intrinsically better people. Much of our recent ideology reflects this implied superiority, from the sanctimonies of political correctness to the metaphysical vanities of contemporary architecture.

The Long Emergency will mark a sharp discontinuity in the circumstances that bred the myth of the perfectibility of man, and all the ideas that grew out of it. Falling standards of living, loss of amenity, shrinking life expectancy, resource scarcity, political disorder, military strife will present a compelling new set of circumstances that will shatter many of our cherished beliefs. I'm not convinced that this will be entirely a good thing. A world moving toward socioeconomic darkness is likely to inspire a lot of ideological and behavioral darkness. Many babies may be thrown out with the bathwater. Personally, I'm comfortable with at least some of the ideas elaborated during the euphoric centuries of the industrial pageant—due process of law, separation of church and state, social equality, the secret ballot, and compound interest, to name just a few. I think we will miss them if they fade away.

Of course, even within the context of recent decades, ideas have shifted dramatically. For example, in the 1950s and 1960s of my youth, gambling was considered a vice, with criminal sanctions applied to it, occupying the distant margins of society. Now, forty years later, gambling is a mainstream recreation in an entertainment-saturated culture. Following that a little further, though, one can't fail to see how the new attitude toward gambling reflects a deeper fundamental shift in normative thinking—that so much current behavior is predicated on the belief that it is possible to get something for nothing.

Forty years ago, the consensus among adult Americans was that it was generally not possible to get something for nothing, and probably harmful for anyone to expect it, try it, or become accustomed to it. Today, with the Las Vegas-ization of our culture, getting something for nothing is a normal condition of life, something to be expected (or at least prayed for). This attitude ends up infecting virtually every other activity in our everyday world, from students who expect to be given automatic As just for showing up, to corporate CEOs who use their companies' operating budgets as their own piggy banks, to ordinary citizens living wildly beyond their means on credit cards.

The particular example I've chosen—getting something for nothing—may illustrate something else, though: the unacknowledged collective drift in consciousness among a people who were once confident about progress and the role of honest effort in it, to a people now utterly cynical about progress and simply wishing for unearned beneficial outcomes in the absence of faith in honest efforts. These changes in collective thinking seem to anticipate the trauma of the Long Emergency now bearing down on us.

The circumstances of the Long Emergency will be the opposite of what we currently experience. There will be hunger instead of plenty, cold where there was once warmth, effort where there was once leisure, sickness where there was health, and violence where there was peace. We will have to adjust our attitudes, values, and ideas to accommodate these new circumstances and we may not recognize the people we will soon become or the people we once were. In a world where sheer survival dominates all other concerns, a tragic view of life is apt to reassert itself. This is another way of saying that we will become keenly aware of the limitations of human nature in general and its relation to ubiquitous mortality in particular. Life will get much more real. The dilettantish luxury of relativism will be forgotten in the boneyards of the future. Irony, hipness, cutting-edge coolness will seem either quaint or utterly inexplicable to people struggling to produce enough food to get through the winter. In the Long Emergency, nobody will get anything for nothing.

I believe these hardships will prompt a return to religious practice in all regions of America, with tendencies toward extremism that will be worse in some places than in others. In the absence of legitimate or effective secular authority, church authority may take its place, perhaps for a long time to come. People desperate for legitimate authority to assist them in organizing their survival will probably accept more starkly hierarchical social relations in general and disdain democracy as a waste of effort. They will be easily led and easily pushed around. This, along with the emergence of a substantial agricultural laboring class, suggests that the ranks of society will be much more distinct in the Long Emergency, with far less movement between the ranks. Do not expect more social equality—expect much less.

Norms of personal conduct may change drastically. Standards of morality will replace the cant of therapeutics. We will be uninterested in the "root causes" of misbehavior and expeditious in dealing with the sheer fact of it, meaning justice is likely to be harsh and swift. Quite a bit of injustice may be a by-product of that, including the persecution of groups and individuals by authorities seeking to impose order at nearly any cost. We will be a lot less inclined to entertain excuses for anything. Personal responsibility will be unavoidable, perhaps excessive. Adolescence as we have known it could disappear and childhood will afford fewer special protections. Reestablished traditional divisions of labor may undo many of the putative victories of the feminist revolution. In the context of new circumstances, these altered relations will come to seem normal and inevitable.

These are daunting and even dreadful prospects. If there is any positive side to the stark changes coming our way, it may be in the benefits of close communal relations, of having to really work intimately (and physically) with our neighbors, to be part of an enterprise that really matters, and to be fully engaged in meaningful social enactments instead of being merely entertained to avoid boredom. The idea of beauty will surely return from its modernist exile, as one of the few consolations in the years ahead will be our ability to consciously craft things for reasons other than to merely shock and astonish. I believe that cases of what we label "clinical depression," in our effort to medicalize all aspects of the human condition, will be steeply reduced, despite universal hardship. When we hear singing at all, we will hear ourselves, and we will sing with our whole hearts.

My Long Emergency

It's not as though I consider myself a detached spectator of all the things I have described, some of them rather terrible to contemplate. I have reasonable expectations to live through the early decades of these epochal changes, and perhaps to suffer because of them. Unfortunately for me, I will not be young. I'm aware of having already lived more than a half century through the greatest fiesta of luxury, comfort, and leisure that the world has ever known. I enjoyed central heating, air conditioning, cheap air fares, cable TV, advanced orthopedic surgery, and computers. It would be churlish for me now to complain about any future hardship. I was less than entirely thrilled by what my culture managed to make of all the advantages conferred by cheap oil, and I made a career of criticizing our behavior in books. I didn't get rich but I supported myself without having to suck up to any boss. I was free and unfettered, and I was grateful to be here at all.

Thirty years ago as a young newspaper reporter, I went through the OPEC oil embargo. The strangeness of it scared and thrilled me, and when it ended after a couple of months I was sure America hadn't seen the end of troubles over oil. I used to drive home from the office up the busy commuter corridor of I-87 out of New York's capital, Albany, and I would marvel at the number of headlights shining at me in the southbound lane and imagine all the engines running behind them, and all the burning gasoline that represented. I would reflect on how similar gluts of commuter traffic poured out of every city, big and small, around the country and how it was just inconceivable that there would always be enough petroleum to run this operation. It would have to end someday.

A few years after the OPEC embargo, I decided to bag my career as a corporate journalist and hunker down in a small town in upstate New York to write books on my own. It worked out all right. I published thirteen books and made a place for myself in an agreeably scaled main-street small-town community, had a social life, got married a few times, and was privileged to travel far and wide around the country and even abroad while pursuing my writing projects. I picked the place I live in for a reason. I feel confident that I am in a good place. It will hardly be untouched by the Long Emergency, but we are surrounded by excellent farmland here and I think my little corner of upstate New York may remain generally civilized.

While I personally believe we are in for extraordinary hard times, I have not taken extraordinary measures to prepare for it, other than choosing to live in a small town in the northeast region of the country. I'm not a survivalist. I'm not hoarding wheatberries and Power Bars in the basement. I own a Leatherman multitool and a Swiss army knife (with corkscrew). I have some woodworking hand tools that don't need to be plugged in. I know how to sharpen them. I have some acquired skill in using them. I have an old sixteen-gauge double-barreled side-by-side shotgun and a few leftover boxes of No. 9 birdshot left over from the days when I bashed around the poplar scrub hunting partridge (not an easy thing to do, by the way). I doubt that I will ever use it in night maneuvers against any host of marauders but, as a character in one of my own short stories once said, "A piece in the closet is a great comfort."

I own an automobile, a pickup truck to be precise, but I don't use it every day. I expect it to be the last automobile I will ever buy. It is fourteen years old and it has 85,000 miles on it; many Toyota engines run 200,000 miles. I expect to do less rather than more motoring in the years ahead and the last thing I want to do is make a large investment in future driving. Eight months of the year, I get around town on a bicycle. In the dead of winter, I caper about on foot.

I'm not so optimistic that the book publishing industry will remain operating the way it has very much further into the future. Way too many books are published these days, and the American public is reading measurably less than ever. I have been publishing a monthly newsletter in my town for about four years as a sort of pro bono exercise. Our daily newspaper, owned by a chain run by chiseling imbeciles, delivers almost no commentary on local affairs, especially on the civic design and development issues that have preoccupied me for the past decade. If the book industry implodes, or I can't make a living in what remains of it, I am determined to run a local weekly newspaper here in my town. It's consistent with my convictions that we are going to have to live locally, and must find something to do that makes us genuinely useful to our fellow citizens—and perhaps incidentally make a living from it.

These days I often find myself ruminating on the twentieth century we have so recently left behind and how strange, awful, and magical it was. I lost my own parents, both within a forty-eight-hour period in the spring of 2001 (quite a coincidence, as they hadn't lived together since 1957). But both of them made it past eighty. I consider it a good thing that they made their exits before the 9/11 attacks, which would have demoralized them intensely at the end, especially my mother, who spent most of her waking hours in old age railing at Cable News Network on TV. They lived through most of the twentieth century, and it must have been quite a romp. They were creatures of the high tide of industrialism and all the residual wonder and faith about the perfectibility of man that managed to survive the two world wars. I doubt they ever gave a moment's thought to the idea of fossil fuel depletion. For them, future shock was the normal condition of life. The automobile came on the scene when they were children. They were already parents before the first spade of dirt was turned for the interstate highway system. They were about my age when Apollo 11 landed on the moon.

I can't think of my parents without thinking, once again, of the music of George Gershwin. My mother, for all her extreme narcissism late in life, was a paradigmatic Gershwin girl in her youth, a snappy, clever babe who smoked cigarettes when it still seemed okay, read *Look Homeward, Angel*, had a lot of romances and a number of marriages (my dad was number 2), and went to Broadway shows back in the days when you could just show up at the box office on a Friday night and get good seats for $1.50. Gershwin was the soundtrack to her life, with all his tenderness and intelligence and the feeling he conveyed musically that everything would turn out all right in the end. In the twentieth century, a lot of things did turn out right for my parents' generation. They beat Hitler, got rich, and lived a long time.

My father was perhaps a little more drawn to Gershwin's blue side. When we all lived together in the 1950s, he liked to sit at the piano and sing quietly, and he often played "Our Love Is Here to Stay," which, sadly, turned out not to be the case with him and Mom. The lyrics still haunt me now that we have left the twentieth century behind:

> The radio, and the telephone,
> And the movies that we know
> May just be passing fancies, and in time may go . . .

How many other familiar things in time may go? What will abide in our collective memory? I don't have children of my own, but I certainly wonder about the world that my friends' children are growing up into, and their children. Perhaps centuries from now when palmettos sprout in the ruined parapets of the RCA building, their descendents will play Gershwin by lamplight on homemade banjos. Perhaps a thousand years from now some monk will crack the code of our all-but-forgotten musical notation and sit down to play "Someone to Watch Over Me" on his ocarina—and something new and exciting will begin. Surely by then, God will have blessed himself back into existence. The human condition is such a mystery. If we don't quite know where we are going, at least we have the satisfaction of knowing where we've been. Sometimes, that seems like enough.

The End

EPILOGUE

In the fall of 2005, after this book had been out for a while, college kids in lecture audiences inevitably asked me when the long emergency might really get under way. "What's your time frame on all this?" is how they would put it. And I would generally answer, "I think we are *in the zone*." I wasn't trying to be cute. Things were definitely happening.

Back in August, Hurricane Katrina roared out of the Caribbean and virtually destroyed modern New Orleans. Not much remained of the city except the old French Quarter beloved by tourists, the mansion-filled Garden District beloved by natives, and the adjacent Uptown neighborhood—and only because the original settlers of this lowland had first built the town on embankments a few feet higher than the rest of the swampy wedge between Lake Ponchartrain and the Mississippi River. With Katrina, all the common supports of civilized existence went down and stayed down: electricity, telephones, sanitation, drinking water, public safety, and food distribution. The behavior of those left behind in the flood seemed as desperate as their circumstances, while government at all levels floundered spectacularly. Months later, next to nothing has begun to be resolved about the reconstruction of New Orleans, in particular the inadequate levees, which, in their current state, could not protect the city against even a less powerful hurricane.

Katrina was equally cruel to the Gulf Coast towns east of New Orleans. It is no exaggeration to say Gulfport was practically wiped off the map. Biloxi, Slidell, Waveland, Bay St. Louis were creamed. Katrina sent her tidal surge miles inland, too, drowning town after town. Then the eye came ashore, and the storm blew a swath of destruction up into Jackson, Mississippi, and as far north as Tennessee. Out in the warm Gulf waters, unseen by the watchers of cable TV, scores of oil and gas production platforms were mangled or collapsed. Refineries and oil platform servicing depots along the coast got clobbered as well, and even pipelines on the seafloor were torn apart. Then Hurricane Rita blew in a few weeks later and smashed up the other side of the big pod of oil and gas rigs that stands concentrated off the Texas–Louisiana coast.

As a result of this one-two punch, 85 percent of oil production and 95 percent of natural gas coming out of the Gulf of Mexico was shut down (or *shut in*, as they say in the oil business), and refining capacity was substantially crippled. The cumulatively shut-in oil production from August to Christmas of 2005 was 18.5 percent of the yearly oil production out of the Gulf and 14.4 percent of the natural gas. Production would not be fully restored to pre-Katrina levels, the industry said, until 2007, and there was the hurricane season of 2006 to consider.

The price of fossil fuels was fibrillating strangely through the fall of 2005. Oil had been edging into the $60-a-barrel range before the hurricanes. The price jumped up to $71 just after the storms. With fewer refineries operating, gasoline prices breeched the $3-a-gallon barrier during the Labor Day holiday, and there were severe distribution problems in the Southeast, leaving angry motorists wilting in gas station lines in Atlanta and Raleigh. But the United States had an agreement with the European Union about this kind of emergency, and, before too long, we were getting roughly 2 million barrels of oil a day from its Strategic Petroleum Supply (SPR), of which about 40 percent was refined product—that is, gasoline, diesel, heating oil, and aviation fuel. This was on top of releases from our own SPR. Throughout the fall, those strategic reserves took on the role of being the world's *swing producer*.

The swing producer is traditionally the entity that can put more oil on the global market by producing extra barrels at will to drive down the price. You might think that oil producers would benefit unequivocally from ever higher prices, but the hard lesson learned over the 150-year existence of the industry is that reasonable price stability has greater value than price jacking because price-jacked economies tend to fall into recessions, cutting demand for oil. So, the swing producer has played an important stabilizing role.

For most of the industry's history, that role was enjoyed by the United States—until our production peaked in 1970. After that, the swing producer was OPEC, in particular OPEC's lead dog, Saudi Arabia. But in recent years, Saudi Arabia has failed to increase production, despite continual pledges to do so, and there was reason to believe that the kingdom was peaking in the fall of 2005. Weird things were happening in its fifty-year-old super-giant Ghawar field—the largest oil field ever discovered and the field that accounted for more than half of Saudi Arabia's total annual production. The Saudis had been resorting to some "enhanced recovery" methods to make up for the loss of natural pressure in the old fields, including pumping massive amounts of seawater down the bore holes, an accepted technique with tricky side effects for the geological structure underground. Now, in 2005, some of the pumps were pulling up 90 percent seawater. It was alarming.

Matthew Simmons, America's leading investment banker to the oil industry, was so concerned about the Saudi situation that he wrote a book about it, *Twilight in the Desert*, published a few months after the hardcover edition of this one. In it, Simmons theorized that the Saudis' giant fields had probably arrived at their own peak. The information had been hard to get because the Saudis regarded their production and reserve numbers as state secrets. Simmons pieced together the puzzle by laboriously studying scores of recondite engineering reports that nobody had bothered to draw conclusions from before. The Saudis, of course, had several reasons to conceal and obfuscate their numbers, including the fact that they were allowed to sell more oil under OPEC quotas if they reported higher reserves (oil left in the ground). But they also had their geopolitical prestige to worry about, while the royal Saud family knew that its grip on the government depended on the nation's oil wealth. In August 2005, the long-ailing King Fahd passed away and was succeeded by his eighty-one-year-old half brother Prince Abdullah. The transition was orderly, and Abdullah seemed to succeed in quashing attacks by revolutionaries on foreign workers.

But by the fall of 2005, the Saudis could not significantly boost oil production, no matter what they tried. Whatever else that meant, Saudi Arabia was no longer able to play the role of swing producer. The markets were now at the mercy of geology, weather, and geopolitics. There was no *governor* on the market machinery. Meanwhile, the world's three other super-giant oil fields producing over 1 million barrels a day—Cantarell in Mexico, Burgan in Kuwait, and Daqing in China—all fell demonstrably into decline. World oil demand, at 82 million barrels a day, was about even with world production. Demand would continue to increase after 2005, but production might never exceed current levels. That was the essence of the peak predicament.

In 2005, a few other things were becoming clear from the fog of assumption and superstition that had shrouded the oil industry for decades before the specter of Peak Oil had wrested the issue from the industry PR shills and their government protectors all over the world and provoked a discussion in the public arena, especially on the free-for-all Internet. One was the apparent fact that much-heralded *new technology* not only was failing to increase total annual global production—which was nearly flat in 2005—but also was now understood to have the ominous effect of depleting the existing oil fields more efficiently. Some of the North Sea fields belonging to Britain, for instance, were showing depletion arcs in 2005 of a staggering 50 percent—meaning that in a few years they would produce nothing. The diminishing returns of technology were finally making an impression.

The Brits were especially worried as they awoke from the raptures of their quarter-century cheap-energy binge, thanks to the aforementioned North Sea bonanza. There was talk entering the winter of 2005–2006—and I write a few days after Christmas—that British industry would have to go to a three-day workweek because of natural gas shortages, so that Queen Elizabeth's subjects would not freeze in their new suburban "estate" homes. For as far ahead as anyone could now see, the Brits would be increasingly dependent on Russia for natural gas imports. This all seemed to catch the New Labor government of Tony Blair by surprise.

The United States had an intense interest in halting the rise of energy prices, for reasons beyond the obvious ones. As well as being hurricane season, autumn was the season for debacles in the finance markets, and since the U.S. economy had become virtually synonymous with the housing bubble (i.e., the suburban-sprawl industry), the prospect for continuing this kind of economic behavior seemed rather ominous. The housing bubble was the suburban home-building industry on steroids. Since the NASDAQ crash of 2001, houses had eclipsed paper securities as Americans' favorite investment medium. *The Economist* reported that 23 percent of the units built in 2005 were bought on pure speculation and were unlived in. Another 13 percent were bought as "second homes." From 2001 through 2005, consumer spending and residential construction together accounted for 90 percent of the total growth in GDP (gross domestic product). And over two-fifths of all private-sector jobs created since 2001 were in housing-related sectors, such as construction, real estate, and mortgage brokering.

Mortgage lending was out of control. The traditional mortgage of 20 percent down with a fixed interest rate on the balance was a relic of the past, replaced by "creative" loans that disregarded any notion of risk. So-called sub-prime loans, made to deadbeats and persons with spotty records of creditworthiness, now made up over 50 percent of mortgages. Norms and standards for lending had vanished in the frenzy to pawn off billets of bundled debt onto a reckless bond market where poor judgment was magically converted into tradable "instruments" and those capital flows fed into other dicey investment sectors, such as the derivatives trade.

The houses themselves, going up in the farthest asteroid belts of suburbia, were on steroids, too. The home-building industry had gotten into the habit of building ever larger ones, for the same reason that the car industry liked to sell huge SUVs: because the profit margin per pumped-up unit was much greater. The average size of a new house had gone up by 20 percent since 1987. Yawning atriums, *lawyer foyers*, and other forms of unusable pretentious space had become standard, home theaters were not unusual, and the master suite mutated into the master *resort*. In the fall of 2005, sliding into the heating season, with the price of gas higher than ever, the suburban builders were stuck with inventories of houses that even well-off professionals might now be reluctant to buy.

Around Halloween there were signs that the years-long housing bull market was cooling. *The Boston Globe* reported a 20 percent fall in prices paid below prices listed for high-end houses. The numbers for November were also down in some of the nation's real estate hot spots. The big national production builders like Pulte and KB Homes took hits on their stock. Pulte dropped the price of its Las Vegas ranchburger tract houses

25 percent in 2005. Anyway, the price of housing in most of the hot markets, such as the San Francisco Bay area, had lost any historically comprehensible relationship to salaries or rents. In Miami, sixty thousand condominium units were either under construction or emerging from the permit approval process. An estimated 70 percent of the buyers there were "flippers" or speculators, and those who had bought with creative interest-only or adjustable-rate mortgages stood to be reamed if the market cooled and they could not expedite their flips. Property owners tended to stave off default as long as possible, and the banks often went an extra mile to accommodate them. It might be spring of 2006 before a picture of the full scope of the housing bubble debacle began to emerge. By Christmas of 2005, the financial markets hadn't cratered, despite tremors such as the September bankruptcy filing of Delta Airlines; the demise of Delphi, General Motors' chief supplier of auto parts; and the designation of GM bonds to "junk" status. Altogether, the stock market showed a minuscule loss for 2005.

Apart from the hurricane extravaganza, it had been a mild fall in the United States, weather-wise. By Thanksgiving the price of oil sank back into the high $50s, thanks to those coordinated releases of strategic petroleum reserves from Europe. In early December it ratcheted back above $60 as the SPR releases stopped and the oil already en route worked its way through the distribution system. This was about the same price it had been just before Katrina struck in late August, which seemed a little odd, given the many disruptions, except for another development little noticed by all but the oil-market insiders: Heavy sour crude was replacing production of light sweet crude as a percentage of American imports at an impressive rate. Much of the build in U.S. crude-oil inventories in 2005 may have been heavy sour crude (apparently, no one tracks inventories of light sweet crude versus inventories of heavy sour crude). Heavy sour crude is harder and more expensive to refine and yields less gasoline than light sweet, so it is worth less per barrel. Some grades of heavy sour crude oil sold for as much as $17 per barrel less than light sweet crude oil in 2005, which basically reflected a shortage of light sweet crude oil. In any case, the average price of oil in 2005, $56 a barrel, ended up 40 percent above the price of oil in 2004.

Natural gas, meanwhile, went up altogether 83 percent above its 2004 price in 2005. America could not necessarily import more of the stuff at will. We were limited to four off-loading terminals for imported liquid natural gas (LNG), and there was no prospect whatsoever of building any new ones quickly, given the complexity and expense of the job. In the fall, with the heating season coming on and gas production still shut in from the Gulf of Mexico, fear briefly drove the price as high as $17 a unit (1 million BTUs). A one-day price jump in November exceeded the absolute price paid at the wellhead in 2002. At $15 a unit, natural gas would be the energy equivalent of oil at $90 a barrel, so the price was veering out of a certain rational economic range (though oil and gas are neither priced on identical criteria nor interchangeable in use).

By Christmas week 2005, which included mild temperatures 21 percent above normal across the United States, gas prices on the NYMEX spot markets had crashed down to the $11 range, which was still 400 percent above the 2002 price. Then just after Christmas, gas shortages across the Atlantic, caused in part by a price dispute between Russia and Ukraine, sent a tremor of fear through western Europe. Russia's Gazprom halted delivery through major pipelines that ran through Ukraine. Soon after, the pipeline pressure began dropping. Supplies to Italy fell 24 percent in twenty-four hours. France saw its deliveries drop between 25 and 30 percent. The result was a price spike for LNG from elsewhere, to such an extent that U.S. LNG companies were hopelessly outbid by the Europeans, and American LNG terminals were paradoxically running at half capacity at a time when American supply was barely keeping pace with demand. The European governments responded to the Russian gambit by commanding their electric utilities to switch to other fuels, chiefly oil, which began driving the worldwide price of oil back above $62.

The wild oscillation of prices was itself a sign of noise in the system. And that noise was symptomatic of the system being unstable because nobody really knew what might happen next, in terms of energy production, cold weather, or geopolitics. A lot could go wrong. This lack of certainty was not a good thing for complex economic systems that relied on stable and predictable prices for crucial commodities. One thing was becoming clear: Countries that owned energy resources, like Russia, suddenly had a powerful lever for jerking everybody else around.

Kenneth Deffeyes of Princeton had famously predicted that the global oil production peak would arrive on Thanksgiving Day 2005 (at 2:30 P.M., or something like that). Now Thanksgiving 2005 had come and gone, and nothing seemed especially different. The freeways were still full of Hummers and Ford Explorers. Christmas shoppers had dutifully reported to the malls, brandishing plastic money. Deffeyes's statement was obviously a gag, but the professor was not altogether lacking in seriousness. His prediction implied that Americans would sit happily bundled in the warm security of their large homes, amid the trappings of their accustomed suburban lives, bathed in the affection of their families, without sensing that anything momentous had happened. And in the general sense, he was certainly correct, just as the U.S. oil peak ticked by in 1970 when the Texas Railroad Commission posted a legal notice in the back pages of a Dallas newspaper saying that producers were released from any quota restrictions and could pump to 100 percent of capacity if they liked—which, at the time, got no more attention from the public than a possum run over on Interstate 35.

Climate change or global warming set off new alarms in 2005, and it was becoming clear even to *denialist* right-wing types that this set of problems would now synergize with and ramify our fossil-fuel problems. Warmer ocean temperatures appeared to be generating an unprecedented sequence of hurricanes that repeatedly smashed oil and gas rigs in the Gulf of Mexico, putting both our economy and our geopolitical security in jeopardy—since any loss of American oil production only makes us more dependent on other nations. A child of ten could hardly fail to understand the connection. An adult, say an oil-industry executive, might even wonder whether this was a trend, and if so, whether it would make sense to repeatedly rebuild billion-dollar oil and gas platforms only to see them get blown down the following summer.

The 2005 hurricane season was the most extreme in U.S. history. There were twenty-six named storms, thirteen of which were hurricanes (winds above 75 mph). Three of the storms—Katrina, Rita, and Wilma—reached Category 5 status, the highest level, meaning winds in excess of 155 mph. For a brief period, when Wilma lingered off the Yucatán, the storm was the most intense ever recorded in terms of measurable interior air pressure. By the end of the season, more than twelve hundred Americans had been killed by the storms, and reconstruction costs were estimated at $200 billion.

Elsewhere in 2005, permafrost was melting in a 1-million-square-kilometer region of Siberian peat lands and bogs formed over eleven thousand years ago, leading to releases of vast amounts of methane gas previously trapped in frozen, partially decomposed organic matter. Methane is twenty times more potent as a heat-trapping atmospheric agent than carbon dioxide. NASA satellite photos showed that the arctic ice cap was shrinking at the rate of 9 percent a decade. The 3,000-year-old Ward Ice Shelf off Canada's Ellesmere Island had cracked in 2000 and by 2005 was breaking into smaller pieces. The retreat of glaciers continued worldwide at all latitudes. Spruce bark beetles in Alaska, enjoying warmer weather and an additional reproductive cycle each season, were chewing up millions of acres of forest. The Antarctic continued to shed offshore shelf ice in 2005, while the Antarctic Peninsula, the tail of land that points up toward Tierra del Fuego, was found to be warming at five times the global average. UK prime minister Tony Blair declared climate change to be "the world's greatest environmental challenge." President George W. Bush wouldn't go so far, granting only that it was "a significant long-term issue that we've got to deal with," while declining to allow the United States to join the Kyoto treaty protocols on reducing greenhouse gas emissions.

In 2005, the outskirts of Los Angeles were hit with two feet of snow, while Medina, in the desert of Saudi Arabia, went through the worst flooding in decades. Boston, Massachusetts, established an all-time January snow record. Ski resorts in Germany had to close because of a lack of snow. Severe drought afflicted much of Asia, from India through China and the Philippines. Stretches of the Missouri River fell to the lowest levels ever recorded. Australia had its hottest April on record, along with severe drought. New England went through the third-coldest May ever, while Seattle issued its first-ever heat warning. Spain's crop yields fell by half because of drought, while forest fires raged through the Iberian Peninsula. Somalia received the first snowfall in its history in June. By July, drought and extreme heat had spread from Spain into France, where locusts invaded the department of Aveyron. In Phoenix, Arizona, temperatures exceeded

100 degrees for weeks on end, with fourteen days in July having highs of 110 degrees or more. Twenty-one people died. At least two hundred heat records were broken that month in different parts of the country. Record rains left parts of Mumbai, India, under water and killed 140 people. Tornadoes swept through Britain, France, and Germany. Then, of course, there were the hurricanes. While Wilma was tearing South Florida apart, New York established the rainiest October ever recorded.

By the very end of 2005, there had not been a major outbreak (pandemic) of influenza in the world, especially the avian flu designated H5N1. But there were worrisome signs. Human deaths from H5N1were showing up here and there around Asia by year's end: thirteen confirmed in Thailand, nine in Indonesia, seven in China, five in Vietnam. There was no evidence yet that the virus had "jumped species." That is, transmission still appeared to occur on a bird-to-human basis rather than a human-to-human basis. Infected wildlife were turning up as far west as Romania.

The overall rate of AIDS infection increased around the world, while rates fell somewhat in individual countries. Kenya and Zimbabwe showed declines of a few percentage points. At the same time, infection rates increased by 25 percent in Central Asia, according to the World Health Organization (WHO). Infection rates were also rising in eastern Europe and South America.

The two biggest geopolitical shocks that occurred after the hardcover publication of this book in 2005 were the subway bombings in London and the riots of Muslim youth gangs in France. In that first edition of *The Long Emergency*, I wrote: ". . . given Britain's large and sometimes overtly belligerent Muslim population, it is something of a miracle—or perhaps a tribute to the British intelligence services—that a major terrorist incident has not occurred there since the 2003 Iraq invasion." Evidently, security has its limits and the Brits' luck ran out. At rush hour on Thursday, July 7, three suicide bombers blew themselves up in separate central London tube stations and a fourth on a double-decker bus, killing fifty-six people and wounding seven hundred. The perpetrators, of course, killed themselves in the process. The terrorist organization al Qaeda quickly claimed responsibility. Three of the bombers were children of Pakistani immigrants, and one was a Jamaican. Two weeks later, on July 21, a second subway bombing plot fizzled when detonators went off but the bombs failed to explode. Nobody was killed. Four prime suspects and several accomplices were arrested in the weeks that followed, and there were no bombings afterward. Months later, an antiterrorism bill was still working its way through Parliament. The bill would chiefly curtail seditious practices that had been previously tolerated, such as preaching the violent overthrow of British society in mosques.

In late October, the housing projects outside central Paris, inhabited mainly by poor Muslim immigrant families, exploded in violence. The event was triggered by the accidental electrocution of two Muslim teenagers who had been hiding from police in an electrical substation. The violence spread all over France, to more than 250 places. It lasted twenty nights. Over a hundred policemen were injured, twenty-eight hundred rioters were arrested, and almost nine thousand automobiles were torched. The monetary damage to property was in the vicinity of $180 million. A nationwide state of emergency was still in effect at this writing.

Elsewhere, the U.S. project in Iraq was still very much under way throughout 2005. The total of Americans killed there stood at 2,185 by year's end. The number killed since President Bush's "mission accomplished" stunt in 2003 had come to equal the number killed before. Roughly 16,000 Americans had been wounded. Medical advances were saving many who would have died in earlier wars. The trade-off was that they survived with more severe injuries. The BBC reported that unofficial estimates of Iraqi civilian deaths varied from 10,000 to over 37,000.

The political picture there remained open to interpretation. Horrible violence occurred virtually on a daily basis all year long. With plenty of help, and a lot of quarreling between sectarian interests, the Iraq interim government cooked up a constitution that was ratified in October after a summer of delays and near breakdowns in negotiations between the three main factions: Sunni, Shi'ite, and Kurd. The election for seats in the new national assembly was held without incident on December 15, 2005, but the results were still far from complete just before New Year's Day 2006.

None of these admirable nation-building steps could obscure the more compelling realities of the situation. One was that our essential reason for being there—namely, to operate a substantial military garrison in the place geographically most favorable for moderating and influencing the behavior of the other troublesome Middle East nations: Iran to the northeast, Saudi Arabia to the southwest, and Syria due west—had not changed one bit in three years. After all, the true mission in Iraq was to guarantee our access to Middle East oil, which amounted to two-thirds of the world's remaining oil; it was reasonable to suppose that our access would be compromised in a region beset by Islamic revolution, sectarian civil war, and other species of mischief.

The nation-building project in Iraq was therefore partly a cover story and partly a practical effort aimed at pacifying the region. We couldn't just appoint some guy to run the place, some new-and-improved strongman, another potential Saddam Hussein. That would have played poorly in the court of world opinion. And the fractured ethnoreligious nature of the country presented obvious difficulties. So the only other choice was to engineer some way for the Iraqis themselves to choose their leaders and representatives, and we decided to pull out all the stops in the process for accomplishing that. Interim government. Constitution. Elections. The works. I'm sure there was some quasi- or crypto-idealistic hope fluttering in the breasts of the American neoconservative strategists that a durable democratic government would come out of this, and that it would act as a *beacon of hope* to the rest of an Islamic world beset by tyrants and mullahs. But mainly, we wanted to have in place a friendly elected government that would invite us formally and *legitimately* to stick around, that is, to maintain military bases there, so we could politely monitor the doings around the region and act as a kind of benevolent police force there. That, I believe, was the core theory—if you sweep away all the extraneous White House and Defense Department PR bullshit and even the sentimental cant of the antiwar movement clouding the issue.

Of course, this desperate mission to use Iraq as a base for pacifying the Middle East assumed that we could actually influence the behavior of these other nations, and whether that was true was not exactly clear. Next door in Saudi Arabia beheadings of kidnapped foreign workers had ceased—perhaps in recognition of the prospect that foreign technicians were indispensable to the Saudi oil industry, and cutting off any more of their heads would have certainly prompted a rush to the exits. The new King Abdullah seemed to be firmly in control, but he was also pretty long in the tooth at eighty-two, and the internal struggles among the many princes vying for power behind the throne were hidden from the news media. Nor were the media kept informed about Jihadist activity in the kingdom. It was probably safe to assume that repression was severe, and that whatever kept the oil flowing was okay with the White House.

On the other side of Iraq, Mahmoud Ahmadinejad, a leader among the "student radicals" who kidnapped American embassy officials in 1979, was elected president of Iran in August and spent much of the rest of the year fulminating at the outside world, especially Israel. In December, for instance, he suggested that the entire population of Israel be removed to Alaska. Through 2005, Iran had maintained its intention to go ahead with a civilian nuclear energy program, and the United States kept making noises of disapproval without doing anything about it. At end of 2005, we were all pretty much back where we started, only with rumors flying around the world's news services that the United States and Israel were planning air strikes in early 2006.

North Korea had been pushed so far to the back burner of world affairs in the second half of 2005 that the doings of Kim Jong Il seemed as remote and forgotten as the doings of the Choson dynasty. Perhaps the world had China to thank for that. The Taiwan question, like that of Korea and its feared nukes, had also magically receded into the globalist gloaming. China did everything possible not to rock the global boat in 2005. It worked sedulously at ramping up industrial enterprise. Its economic growth was reported by *The Economist* to be a hypertrophic 9.3 percent. It was reaching out aggressively to forge contracts for Canadian and Venezuelan oil and to build a pipeline across Kazakhstan, but its status as net energy importer remained rather ominous. China also faced horrendous environmental problems and was governed by a weird crypto-Communist bureaucracy that, despite all the trappings of free enterprise, ruled by force and violence.

At home, despite the hurricanes, the fibrillations of the oil and gas markets, and our vicissitudes in Iraq, the public still apparently subscribed to the notion that the American way of life was nonnegotiable. The ceaseless pursuit of cheeseburgers and chain-store bargains went on unabated, while from sea to shining sea the bulldozers scoured the meadows, cornfields, and sagebrush scrub to plant millions of new McHouses in the farthest reaches of the metroplexes. No figure of standing in either political party would dare inform the public that an economy based on a suburban-sprawl housing bubble did not have much of a future in the energy-challenged world we were obviously entering.

A parallel idea seemed to prevail across the nation—and across the political spectrum—that there was no need to worry about oil and gas because some other fuel or system or *technology* would come along, at *the market's* coy bidding, to save our asses. This wish-based faith turned up in the strangest places. For instance, I gave a talk at the headquarters of the Google company in California's fabled Silicon Valley. It was a bizarre experience (for me; I can't speak for them). The Google building, situated in a suburban office park, was tricked out like Pleasure Island in the Disney version of *Pinocchio*. There were snack stations loaded with gummy creatures, granola, and malted milk balls every fifty paces, and all kinds of recreational equipment deployed around the place: pool tables; knock hockey; foosball; video games; and vibrating massage chairs. They had a very nice auditorium, where I gave my spiel on Long Emergency–related issues. Half the audience of Google employees were dressed like skateboard rats. No doubt some of them were millionaires. During the questions-and-comments period that followed my blab, several indignant audience members rose from their seats and said, in effect, "Like, dude. We've got . . . like . . . technology!" They couldn't buy the idea that we could get blindsided by a permanent global energy crisis. And they had been so successful—become so rich—moving little pixels around video screens that they seemed to assume that the energy rescue remedy was just a few mouse clicks away. I couldn't persuade them that life is tragic—in the sense that happy endings really aren't guaranteed, not even for Americans.

I think the year ahead will be much richer in events than the year past. If global oil production does stay flat, as it did in 2005, or even contract, as it might, we will have the first glimpse in the fabled rearview mirror at the awesome global oil peak. In the face of it, I don't know what it will take to mobilize our political leaders to recognize the need for a new American Dream, and to comprehend the tremendous effort it will take for us to get there.

The Republican party enters 2006 in a three-ring circus of scandal that will probably galvanize the public's attention the way Watergate did in 1973. As someone who was a young newspaper reporter during the Watergate scandal, I readily admit that it was a lot of fun and eventually very satisfying, with old Nixon waving farewell from the steps of Gerald Ford's helicopter. For all of its dramatic satisfactions, though, I can see now how it distracted the nation from the first great energy crisis: the OPEC embargo. We didn't pay sufficient attention to really get the message—which was that we needed to make plans to live differently. The Iran hostage crisis distracted us from the second oil crisis. Now, here we are, on the verge of congressional hearings and court proceedings against Mr. Abramoff and the politicians he paid off, and the officials who spied on Americans without warrants from the courts, and the figures implicated in the Valerie Plame affair, and the alleged campaign contribution violations of the former house Republican leader, Tom DeLay, and the alleged stock trading irregularities of the senate majority leader, Bill Frist . . . and, well, you see how it goes.

In his 2006 State of the Union speech, President Bush only feebly addressed the nation's energy predicament (and I add this now just as the epilogue goes to press February 1). Bush admitted that "America is addicted to oil," but proposed a set of techno-remedies that have become a joke in serious circles: bio-fuels that require greater "inputs" of energy to grow than they give back; hydrogen cars (forget about it); and ethanol, of which one wag remarked, "If oil is our heroin, then ethanol is our methadone."

Like many leaders in government and business who are sunk in denial about what circumstances require of us, Bush let on that we can continue living in a drive-in utopia by just switching fuels. Bush could have seized the moment by proposing to rebuild the national railroad system—a project that would create thousands of good jobs, while benefiting Americans of all ranks and conditions—but he said not word about it. Nor did the Democrats, by the way, in their fatuous response to the president's speech.

As a registered Democrat, I have worries about the failures of my own party, the putative opposition to the neocons now ruling. Why has not one national Democratic politician come forward and proposed to fix the American railroad system—the one project we could certainly address with confidence? Or leveled with the public about our grotesque levels of debt? Or discussed some alternatives to an economy based on speculative real estate? Political parties in America do sometimes vanish in a cloud of irrelevance. In 1852, the Whigs ran America. By 1858, they had been spat out of the universe like so many cosmic watermelon seeds. I hope that doesn't happen to my Democrats, but it's beginning to look like it might.

Politics, like nature, abhors a vacuum. With the Republicans' stock possibly reduced to junk status after the circuses ahead, and the Democrats consumed in fugues of irresolution while pandering idiotically to tired old identity politics, some new political *thing* could rear its head as we enter both the next national election cycle and the first years of the hardships I call the Long Emergency. I am dogged by the thought that Americans have prepared themselves so poorly for what is coming at them, that they will prove to be eager to have someone aggressively direct them toward purposeful activities—in other words, to push them around and tell them what to do. I hope that is not how things shake out. I hope we can remain an autonomous people, within the framework of our pretty good democratic institutions. But it's not an entitlement, any more than life in a 4,000-square-foot vinyl-and-chipboard house thirty-eight miles outside Minneapolis is an entitlement. Perhaps the great political question of the years ahead is: How do we become a reality-based nation?

James Howard Kunstler was born in New York City in 1948. He is the author of three other nonfiction books, *The Geography of Nowhere, Home from Nowhere,* and *The City in Mind: Notes on the Urban Condition,* as well as nine novels, including *Maggie Darling: A Modern Romance, The Halloween Ball,* and *An Embarrassment of Riches*. He has been an editor with *Rolling Stone* and his articles have appeared in *The Atlantic Monthly* and *The New York Times Magazine.*

Afterword

There are two realities "out there" competing for verification among those who think about national affairs and make things happen. The dominant one (let's call it the Status Quo) is that our problems of finance, economy, and energy will self-correct and allow the "consumer" economy to resume in "growth" mode. This view includes the idea that technology will rescue us from our fossil fuel predicament—through "innovation," through the discovery of new liquid fuel rescue remedies, and via "drill, baby, drill" policy. This view assumes an orderly transition through the current "rough patch" into a vibrant reenergized era of "green" Happy Motoring and resumed blue-light special shopping.

The minority reality (let's call it the Long Emergency) says that it is necessary to make radically new arrangements for daily life and rather soon. It says that a campaign to sustain the unsustainable will amount to a tragic waste of our dwindling resources. It says that the "consumer" era of economics is over, that suburbia will lose its value, that the automobile will be a diminishing presence in daily life, that the major systems we've come to rely on will founder, and that the transition between where we are now and where we are going is apt to be tumultuous.

Since the change my view proposes is so severe, it naturally generates exactly the kind of cognitive dissonance that paradoxically reinforces the Status Quo view. The more the broad public senses troubling change in the wind, the more it wishes to preserve all the familiar, comfortable trappings of life as we have known it. The dialectic between the two realities can't be sorted out between the stupid and the bright, or even the selfish and the altruistic. American business is full of MIT–certified, high-achiever Status Quo techno-triumphalists who are convinced that electric cars or diesel-flavored algae excreta will save suburbia, the three thousand mile Caesar salad, and the theme park vacation. The environmental movement, especially at the elite levels in places like Aspen, is full of Harvard graduates who believe that all the drive-in espresso stations in America can be run on a combination of solar and wind power. I quarrel with these people incessantly. It seems especially tragic to me that some of the brightest people I meet are bent on mounting the tragic campaign to sustain the unsustainable in one way or another. But I maintain that life is essentially tragic in the sense that history won't care if we make bad collective decisions or whether we succeed or fail at carrying on the project of civilization.

While the public supposedly voted for "change" in 2008, I sensed that they badly underestimated the changes really at hand. I voted for "change" myself, I suppose, in pulling the lever for Barack Obama, regarding him as a figure of intelligence and sensibility. But I'm far from convinced that he truly sees the kind of change we are in for, and I fret about the measures he'll promote to rescue the Status Quo.

The Early Innings of the Long Emergency

Without reviewing all the vertiginous particulars of the year now ending, suffice it to say that the U.S. economy fell on its ass and that the "global economy" did a face-plant in 2008. The American banking sector imploded spectacularly to the degree that investment banking actually went extinct—as if a meteor landed on the corner of Madison Avenue and 51st Street. The response by our government was to shovel "loans" onto the loading dock of every organization that pretended to be something like a bank, while "bailing out" an ever-longer line of corporate pleaders who came to the door with a pitiable song-and-dance. The oil markets went on a roller coaster ride. The housing bubble collapse geared up to avalanche velocity (taking out whole colonies of realtors, mortgage brokers, and construction contractors in its path), the commercial real estate sector developed hemorrhagic fever, retail drove off a cliff, the stock market lost about 40 percent of its value, jobs and incomes vanished, and tens of millions of ordinary citizens addicted to revolving credit found themselves in a life-and-death struggle for the means of existence. None of this is over yet.

The Year Ahead

Much of what was lost in 2008 will not be recovered: enterprises, personal fortunes, chattels, reputations.

I expect a period of euphoria to mark the early months of the Obama government, apart from whether the new crew can accomplish anything. It will be a relief just to have a president who speaks English correctly and has experienced something like real life prior to politics. Restoring credibility and legitimacy in leadership will be a big deal. If President Obama can bring himself to tell the truth about our situation, we may recover a collective sense of consequence.

A sign of positive "change" might be the commencement of prosecutions for misdeeds in banking and securities that are now destroying the entire system of deployable capital. The time when it was enough to claim that "mistakes were made" is over. A good place to start will be an investigation of Henry Paulson for insider trading stemming from Goldman Sachs's shorting of its own issued mortgage-backed securities when Mr. Paulson was the company's CEO. Beyond his case, there should be enough work at Attorney General Eric Holder's office to employ a line of law school graduates stretching from Brattle Street to the planet Mars. It will be salutary for the nation to see those who engineered the banking collapse come to greater grief than the mere surrender of their Gulfstream jets and Hamptons villas.

By the way, being allergic to conspiracy theories, I don't believe for a minute that there is some kind of shadow elite of "Bilderbergers" standing in the background to protect these grifters – and I also believe the reason these paranoid notions persist is because it is otherwise hard to account for the extravagant irresponsibility of the Bush circle and its servelings.

Apart from "cleaning up Dodge," so to speak, and from issues of collective character-and-conscience-in-office, I worry that the avalanche of troubles already ongoing will overwhelm Mr. Obama and his people. It's also well worth worrying whether they will pursue policies similar in kind to the ones pursued by Bush and Company, namely throwing money at everything and anything, and it sure looks like they are planning to do just that. I am especially concerned about an "infrastructure stimulus" project aimed at highway improvement at the expense of public transit. This would be the epitome of a campaign to sustain the unsustainable. We need to begin planning right away for a transition away from car dependency, not in order to be good socialists but because Happy Motoring is at the core of our unsustainability trap. The car system is going to fail in manifold ways whether we like it or not, and it will fail due to circumstances already underway.

For one thing, it will cease to be democratic as the remnants of the middle class find it impossible to get car loans, or pay for fuel, or insurance, and that will set in motion a very impressive politics of grievance pitting those who are still able to enjoy motoring against those who have been foreclosed from it. Contrary to what you might make of the current situation in the oil markets, we are in for a heap of trouble with both the price and supply of petroleum (more on this below). And there is little chance that any techno rescue remedy to keep all the cars running by other means will materialize.

The stock markets took an historic beating in the fall of 2008. The harsh reality may be sinking in that economic "growth" in the context of industrial expansion is faced with primary resource shortages in the years ahead, and the investments that stand for our hopes and expectations for "growth" may therefore lose value and meaning. I would therefore not be sanguine about any return to booming stock markets in anything but short "dead cat" bounces.

Meanwhile, jobs will evaporate by the millions and companies will go bankrupt by the thousands, especially in the so-called service sector, and in all the suppliers of such, along with the landlords in all the malls and strip malls. The desolation will mount quickly and will be obvious in the empty storefronts and trash-filled parking lagoons. In the event, two things will become increasingly clear to the nation: that the consumer economy is dead, and that there is no more available credit of the kind that Americans are in the habit of enjoying.

We're turning around to discover that we are a much poorer nation than we thought because from now on credit will be extremely hard to get for anyone for any reason. The businesses that survive will have to keep going on the basis of accounts receivable. This is the area where the crash of giants will be heard. I've been saying since the first edition of *The Long Emergency* that comprehensive downscaling in all our activities, from farming to business to schooling to governance, will be the categorical imperative of the years ahead. Giant enterprises requiring giant loans to get from quarter to quarter will tend to not make it. Borrowing from the future will become a practical impossibility as past bad debts from previous borrowings continue to unwind, cease performing, and get written off. This argument implies that the federal government will tend to flounder just as General Motors, Citicorp, Target stores and other gigantic enterprises will tend to flounder.

It would be sad to see President Obama hamstrung as the head of an impotent oversized government that can't seem to do anything right, even with the best intentions. And it is why I see his best role under the circumstances as largely symbolic—as a reassuring presence encouraging the distressed public to bravely bear their hardships, and be kind and helpful among their neighbors.

Households, like businesses, will have to pay as they go from earned income. The house as ATM is over. Credit cards are maxed out and credit ceilings are ratcheting down like the ceiling in "The Pit and the Pendulum," preparing to slice-and-dice the old "normal" of family life in America. Bankruptcy will be the new Nascar. A lot of families will lose everything. They will sift and disperse into the housing owned by other family members—parents, siblings—and a strange new not-altogether-comfortable kind of togetherness will become common. Over time, a lot of people will go looking for casual work "under-the-table" (and probably low-paying). These workers will begin to look and act like a new servant class, and before too long they may be absorbed into the households of people who employ them. There will be plenty of room for them there. The relationships could well be mutually beneficial, not necessarily exploitative.

Municipalities, counties, and states will join in the bankruptcy fiesta. Expect collapsing services as a result, a situation fraught with danger—of rising crime, also of public health emergencies as water systems are not kept up, sewage treatment becomes dicey, EMT and fire prevention staffs are pared down. I don't imagine the federal government stepping into every podunk or metropolis from sea to shining sea and propping up these services. People will have to cope with danger and deprivation.

2009 may turn out to have been the point where we begin to understand what kind of places will be more hospitable to human society further ahead. I maintain that our giant urban metroplexes have way overshot their sustainable scale and will contract severely. With all the economic hardship, we ought to expect a lot of demographic churning, people leaving troubled places—Las Vegas, Phoenix, parts of California, Houston, Atlanta—and moving on to some place more promising. Small towns and the smaller cities, places consistent in scale with the energy realities of the future, will be reactivated. Proximity to productive farmland will be especially important as petro-agribusiness stumbles.

At the time of this writing, the reorganization of the rural landscape into smaller-scaled farms has not begun to occur. The years following the Wall Street meltdown of 2008 are likely to be very hard on agribusiness, given the shortage of capital for an enterprise addicted to revolving credit. There is also plenty of room for more mischief in the oil markets. Eventually, the rural landscape will require the labor of many more people than is currently the case. Whatever else happens, we'll surely see a massive return to home gardening as household budgets become strained to the extreme. As the New Urbanist Andres Duany said recently, "Gardening is the new golf!"

The Oil Scene

Many were stunned in 2008 to witness the fantastic parabolic rise and fall of oil prices up to nearly $150 and then back around $36 by Christmas time. Quite a ride. I said in *The Long Emergency* that volatility would be the hallmark of post peak oil because it was obvious that advanced economies could not absorb super high prices and would crash in response; that at some point after crashing, these economies would respond to the new lower oil price, resume their cheap oil habits, and build to another oil price rise . . . and crash again . . . in a ratcheting declension of industrial output through each cycle of crash and rebound.

What I probably didn't realize at the time was how destructive this cycling between low-high-and-low-again oil prices would actually be in the first instance of it, and what a toll it would take right off the bat. We can see now that our journey through the first cycle effectively destroyed the most fragile of the complex systems we depend on: capital finance. As a result, a huge amount of capital (say $14 trillion) has evaporated out of the system, never to be seen again (and never to be deployed for productive purposes). It will be harder for the USA to rebound from the grievous injury to this crucial part of the overall system in the next cycle. We simply have fewer capital resources to apply to any transition from our old cheap-oil ways to whatever follows. European finance and economies have foundered similarly—though the European nations are not burdened to the same degree by the awful liabilities of suburbia.

Even if these advanced economies—throw in Japan too—remain moribund, the price and supply prospects for oil look ominous. My own guess is that the price of oil overshot on the low end just as it overshot on the high end in mid-2008, and that, when all is said and done, we'll still see an upwardly trending price line over the long haul. The plunge in oil prices was as much the result of banks, hedge funds, and individuals dumping oil investments and positions to raise cash as it was a matter of the markets predicting a sharp fall-off in economic activity, and supposedly oil consumption. The truth is that "demand destruction" for oil in the USA has been surprisingly mild compared to the drop in price. Gasoline consumption dropped from 9.29 million barrels a day in 2007 to 8.99 million barrels a day for 2008. That's not much of a fall-off, especially compared to the price drop.

As Julian Darley of the Post Carbon Institute put it so nicely: "there won't be any energy bailout." As many other people have noted, the plunge in oil prices through early 2009 strongly ensures future *supply* destruction, since so many planned oil projects were suspended or cancelled because they are economic losers at $40-a-barrel (or even $70). Even projects well underway, such as Canadian tar sand production, have been scaled back or shut down because they don't make sense at current prices. Some of these other newer projects will now never get underway—they have missed their window of opportunity with so much capital leaving the system—and so the hope of offsetting very-near-future depletions in old oil fields looks dimmer and dimmer.

Those depletions are very serious. For instance, Mexico's super-giant Cantarell oil field, the second-largest ever discovered after Saudi Arabia's

Ghawar field, has shown a 30 percent depletion rate in 2008 alone. (Pemex had forecast a 15 percent rate entering the year.) Overall, Mexico's export rates declined more than nine percent in 2008. Cantarell provides over 60 percent of Mexico's total production, and Mexico is America's third largest source of imports—just after Saudi Arabia (#2) and Canada (#1). Obviously, Mexico soon will lose its ability to export oil, and as that occurs, America is going to feel more than a pinch—more like a two-by-four upside the head. In short, remorseless depletion is underway and we are less likely now than even a year ago to make up for it. Mexico's problem illustrates more generally the oil export crisis now manifest among all the nations who have had a surplus to sell. One way or another, they are all using more of their own declining oil supplies, and their net exports declines in percentage now exceed their production declines through depletion.

At some point, then, demand, even if slightly lower worldwide, will catch up with declining supply. I sense that we will see two things occur in the near future, possibly at the same time: a resumption of rising oil prices and spot shortages. I say this because the global economic fiasco is already creating geopolitical friction, and inasmuch as America has to import almost three-quarters of the oil we use, the prospect for trouble is great.

The tragic part of all this, of course, is that the temporary plunge in oil prices has prompted an incurious American public to assume, once again, that the global oil predicament is some kind of a fraud. Given the flood tide of fraud they have been subject to in banking and investment matters, I suppose you can't blame them from thinking that everything is some kind of a scam. There are reports lately that an increasing percentage of the few cars sold now are SUVs. How soon we forget.

Though I give Boone Pickens high marks for stepping up to the leadership plate, I'm not altogether onboard with his energy proposal for swapping natural gas for gasoline in motor fuels while we swap out wind power for natural gas in electric power generation. I don't believe that the ballyhooed shale-gas plays of the last few years will prove to be sustainable long-term, as some huckster's claim. They are expensive to drill and run, and they all tend to deplete very quickly – around one year. I'm not convinced we have the capital or the resources even to come up with the steel necessary to drill for it. Anyway, the last thing we need is a way to prolong our car dependency, and the core of Mr. Pickens plan rests on the assumption that car dependency must continue.

Many other people hope strenuously that various alt.energy systems will insure the continuation of Happy Motoring. This is an idle hope. The new era of hoped-for change will be very sobering for those who imagine that hybrid cars, or electric cars, or "air" cars, or natural gas cars, veggie oil cars, or any other kind of car technology will save the day. Even if President Obama mounts an "infrastructure stimulus" program, it will not keep up with all the necessary routine road repair that our highway system requires. The extreme financial hardship faced by localities and states ensures that they will have to postpone a lot of expensive highway maintenance—even if the federal government fixes a big bunch of bridges and tunnels—and so we face the interesting prospect that our roadway systems will enter their own deadly cycle of systemic failure even before the whole car issue is settled. Meanwhile, I am waiting to see whether President Obama will undertake a restoration of passenger railroad service. I've said enough about this in the past, but it's worth reiterating that a failure to get comprehensive passenger rail service going will be a sign of how fundamentally unserious we are as a society.

The Specter of Inflation

This is the "other shoe" that many are waiting to hear drop. Right now we are caught up in a compressive debt deflation as mortgages stop "performing" and loans of all kinds are welshed on. Since money is loaned into existence, and a great many loans are not being repaid, then a lot of money is going out of existence. That's what I mean when I say that capital is leaving the system. At the same time, the Federal Reserve has made good on its promise to drop money from helicopters if necessary to prevent an implosion of the banking system (as all that older, welshed-on money goes out of existence), and so it's now a question as to when the amount of new money will exceed the disappeared old money. In any case, there is bound to be a lag period between the time that the Fed's money is dropped from the choppers and the time it actually filters through the banks and other recipients to the so-called "real economy" of people who buy and sell real things. The credible estimates I hear run between six and eighteen months. Right now, the big banks who have received bail-outs are simply sitting on their money, refusing to lend (because they are fundamentally insolvent).

I'll only venture to guess that we could see the start of serious inflation before the next edition of this book goes to press. To some extent, all currencies are now free-falling together, some at slightly faster rates than others, but the situation of the U.S. dollar is so grotesquely dire, and our structural imbalances of finance so monumental, that it is hard to imagine our currency will not win the international race to the bottom. Gold resumed its movement upward against the dollar a week before Christmas 2008. The government—and anyone badly in debt—benefits much more from inflation than deflation, as under inflation debts grow cheaper to pay back in "money," so every effort will be made to avert the latter. The trouble lies in the government's dumb incapacity to control dangerous things that it sets in motion, so that an inflationary campaign to avoid compressive deflation can so easily lead to a fiasco of high or even hyper inflation—the kind that kills governments and turns societies into murderous machines.

Geopolitics

Well, now, who the hell knows what's in store. Aside from a few bombs here and there, and pirates skulking around the horn of Africa, the world scene was miraculously free of major incidents in 2008 – perhaps the worst being a toss up between the November Mumbai bombings and the earlier fiasco in Georgia, where the U.S. prompted Georgian President Mikheil Saakashvili to send troops into the South Ossetia region and the move was answered by overwhelming force from neighboring Russia, leaving the United States looking feckless and retarded for its troubles. But otherwise, there wasn't a whole lot of action out there.

Until the last few days of 2008, that is when Israel began a strong counteroffensive against Hamas rocket attacks issuing from the Gaza strip. I wonder if it is the beginning of a new coordinated offensive by Islamic extremism aimed at taking advantage of the West's current plight (and the

West's probable aversion to do anything that will complicate its hoped-for economic recovery). Any coordinated campaign might be aimed at confounding the new American president.

The other hot corner of the world as we go to press is the India-Pakistan border where the sixty-year-old rivalry, which has already produced three wars, looks to be gearing up for yet another round. I'm not the first one to say that Pakistan is an extremely dangerous regional player, being an economic basket case, possessing a score or so of nuclear bombs, harboring more Islamic fundamentalist maniacs than any other place in the world, and having a government held together with duct tape and twine. The caper in Mumbai last September could well have been construed as an act of war, but somehow India kept its head. Who knows where this is going?

So far I have only described what is already obviously going on. Add to this that Iran is closer to achieving membership in the atomic-weapon club. They've been spinning their centrifuges without any let-up and nobody has done anything about it. My guess is that neither the United States nor Israel will attempt to take out their facilities anytime soon. Of course, if Iran used a nuclear device against Israel, or anybody else they could reach out and touch with a missile, they would be asking to become the world's largest ashtray. End of story. A different story, though, is how Iran might behave if and when the U.S. military presence in Iraq is reduced. I can imagine Iran doing anything possible surreptitiously to gain control over Iraq's southern oil regions around Basra, but even the Iraqi Shia don't like the Iranian Shia that much. Anyway, Iran's economy has suffered hugely from the fall in oil prices. That nation may be in for more internal trouble than they have seen in thirty years since the Shah was tossed out by the minions of Ayatollah Khomeini.

There's been a lot of sentiment around the blogs and in the conference break-out sessions that as the United States and Europe fall into economic disarray, China would emerge as the great new hegemonic superpower. While it's come a long way in a quarter-century, China's internal problems are still enormous and worsening. They're in trouble with water, food imports, mass unemployment, and energy. They have locked in some oil contracts around the world, but they are still susceptible to vagaries in the oil markets and Black Swan events. As the U.S.

consumer economy falls into a coma, and the shipping containers from China to WalMart get sparser, the Chinese government will face the wrath of millions of unemployed workers. I believe China will struggle desperately, perhaps growing more surly as the U.S. dollar inflates and their holdings of treasury bills begins to look more like a swindle.

Russia may be suffering economically for the moment due to the crash of oil prices, but they are energy-resource rich, and if they don't like the current price, they can keep more of their oil in the ground until the price looks more attractive. I think Mr. Putin has the confidence of the Russian people and will survive the current malaise.

Japan remains a riddle wrapped in toasted *nori*. They're beggaring their own factory workers to stay solvent. Their banking sector has been zombified for a generation. They import 95 percent of the energy they use. Do they have a plan? One can imagine them sliding in resignation back to something like the sixteenth century, giving up the whole industrial circus as more trouble than it's worth, just as they once gave up on firearms.

The over-arching geopolitical theme up ahead will be the end of robust globalism as we've known it for some time. Reduced trade, increased competition for energy resources, sore feelings over debts and currencies will drive nations inward. Note to Tom Friedman: the world turned out to be round after all.

Conclusion

The big theme in the next phase of the Long Emergency will be contraction. The end of the cheap energy era will announce itself as the end of conventional "growth" and the shrinking back of activity, wealth, and populations. Contraction will come as a great shock to a world of conventionally programmed economists. They will toil and sweat to account for it, and they will probably be wrong. This contraction will do its work in unpleasant ways, driving down standards of living, shearing away hopes and expectations for a particular life of comfort, reducing life expectancies, and introducing disorder to so many of the systems we have depended on for so long. People will starve, lose their homes, lose incomes and status, and lose the security of living in peaceful societies. It will become clear that the Long Emergency is underway.

My hope, at least for my own society, is that we will transition away from being a nation of complacent, distracted, over-fed clowns, to become a purposeful and responsible people willing to put their shoulders to the wheel to get some things done.

My motto nowadays: "*no more crybabies*!"